Beyond Webcams

Beyond Webcams

An Introduction to Online Robots

edited by Ken Goldberg and Roland Siegwart

The MIT Press
Cambridge, Massachusetts
London, England

This book was set in Times Roman and Arial Narrow by the authors and was printed and bound in the United States of America.

Library of Congress Cataloging-in-Publication Date

Beyond Webcams : an introduction to online robots / edited by Ken Goldberg and Roland Siegwart.
    p. cm.
   Includes index.
   ISBN 0-262-07225-4 (hc. : alk. paper)
    1. Robots—Control systems. 2. Remote control. I. Goldberg, Ken. II. Siegwart, Roland.

   TJ211.35 .B49 2001
   629.8'92—dc21

2001044325

For links to active and archived online robots, see:

http://mitpress.mit.edu/online-robots

Contents

Part IV, Other Novel Applications

Foreword

Thomas B. Sheridan

It has been just over 50 years since the first "remote-manipulators" were developed for handling radioactive materials. Outstanding pioneers were Raymond Goertz and his colleagues at the Argonne National Laboratory outside of Chicago and Jean Vertut and his engineers at a counterpart nuclear engineering laboratory near Paris. Their first systems allowed operators to stand outside of radioactive "hot cells," peer through leaded glass radiation barriers, and grip "master" arms coupled to "slave" arms and hands inside the cells, which in turn grasped the remote objects. These were all-mechanical devices, where both master and slave had six-degrees-of-freedom arms plus simple gripper hands, and the linkage between master and slave was via cables or metal tapes. Both laboratories were soon applying electromechanical and electrohydraulic servomechanism technology to the actuators, so that the coupling could be via electrical wires rather than mechanical linkages.

In the late 1950s and early 1960s, engineers began dreaming of remote manipulation where the operator and the manipulation task environment were an arbitrary distance apart and visual feedback was via TV. Sputnik provided the incentive for a race into space as well as experimental efforts to build lunar roving vehicles and remote manipulators for the moon. This class of system, whether vehicle or manipulator, came to be called a *teleoperator*— an operator (in the generic sense) at a distance (from the human). It soon became evident that one had to cope with the three-second time delay imposed by the round-trip of radio communication if one wished to control such a vehicle from earth. In the early 1960s Ferrell and this writer at MIT and Thompson at Stanford demonstrated what was involved for a human to perform remote manipulation with time delay. Soon afterward we began pushing the idea of supervisory control where the human communicates a program of instructions to the remote manipulator or vehicle, and the control loop is closed through the sensors and actuators at the remote site. The subclass of teleoperator that then acts autonomously, based on such supervisory or teleprogramming by its human monitor, has come to be called a *telerobot*.

Goertz's lab built the first head-mounted and head-tracked display to give the operator a sense of telepresence (though the term was popularized much later). In my lab Strickler demonstrated in 1966 what I believe was the first "teletouch" system for a teleoperator. In that same decade significant new developments occurred in undersea teleoperators made by American and European navies and by suppliers to the off-shore oil industry. The U.S. Navy employed a teleoperator to recover an accidentally dropped H-bomb in the deep ocean off the Spanish coast.

Enormous strides were made in teleoperators and telerobots during the four decades that followed. The electromechanical devices, including sensors, actuators, computers and communication technology, were all much improved, miniaturized in many cases, and made to withstand the indignities of the factory floor, outer space, and the deep ocean. Much effort was put toward providing the telerobot with elegant software that made it easier for the human to program and monitor it, and made it much more capable when it needed to act on its own.

In recent decades *haptics*, the generic term for force sensing and perception by the human through the skin, muscle, tendon, and joint senses (including touch, kinesthesis or sense of motion, proprioception or sense of body posture), has come into its own. These integrated forms of apprehending our environments, in contrast to natural evolution in which they antedate vision and hearing, are late-comers to man-made technology of vision and hearing (telephonics, radio, photography, and video). At the same time in engineering robotics, impedance control, meaning adaptive. stiffness or compliance, has been an area of considerable analytic and practical development. Teleoperators and telerobots now perform tasks in such varied areas as fruit picking and logging, undersea pipeline inspection, military and police surveillance, sorting cells under microscopes, drilling holes in bones for implantation of artificial joints, and endoscopic surgery.

In the last decade the Internet has blossomed into an engine of communication with staggering implications for both institutions and individuals. Most people, however, continue to think of the Internet as a means of sending e-mail and getting information from remote databases. They remain unaware that another huge class of operations lies just ahead, namely the ability to control physical things remotely over the Internet. What kinds of things? Anything one can imagine a telerobot doing — with telerobot interpeted very broadly. When away from home, for example, one could turn up the heat, start the preparation of a meal, feed the cat, put out the garbage, examine the mail, or check whether Bobby cut the grass. An office or factory manager could inspect the work of others or ready a product for shipping. A student could inspect some ancient ruins in a museum, perform an experiment on the ocean floor, shake hands or participate in an experiment or athletic activity with students in another country.

Goldberg and Siegwart offer us a tantalizing peek into this future, a future of DOING things — physical interactions — over the Internet that few have even considered. Of course there will be problems of access. For some robotic devices access will be rigidly restricted, and the Internet's key advantage is the flexibility of where the operator can gain access to communication. In other cases access will be much more open, even designed to attract the widest possible participation. This raises new questions and problems in democratic control: how might multiple people participate in control of a telerobot? Perhaps experiments in that arena will help us understand how multiple people can participate in controlling any complex system or institution.

Goldberg and Siegwart's excellent array of authors in this book covers many aspects of the problem of controlling physical interactions over the Internet: from science and engineering to education and art, from hardware to software and linguistics, and from outer space to proteins. This book is a first.

Acknowledgments

It has been a true pleasure to work with the generous contributors to this volume and with Doug Sery and the talented staff at MIT Press.

The following is a very incomplete list of the people we would like to thank for sharing their ideas and insights:

Tiffany Shlain, Billy Chen, Eric Paulos, Michael Grey, John Canny, Chris Perkins, Dezhen Song, Michael Mascha, Gerhard Schweitzer, David Pescovitz, Steven Gentner, Illah Nourbakhsh, Jeff Wiegley, Joe Santarromana, Ann Goldberg, Richard Wallace, Peter Lunenfeld, Gerfried Stoker, Henrik Christensen, Rosemary Morris, Francesco Mondada, Roger Malina, Lind Gee, Carl Sutter, Hubert Dreyfus, Adam Jacobs, George Bekey, Erich Berger, Mark Pauline, Judith Donath, Karl Bohringer, Kai Arras, Woj Matusik, Bob Farzin, Leonard Shlain, Rick Wagner, Randall Packer, Ruzena Bajcsy, Rene Hartmann, Antonio Bicchi, Heidi Zuckerman-Jacobsen, Rod Brooks, Nicola Tomatis, Dave Cannon, Brian Carlisle, John Craig, Michael Erdmann, Florian Brody, Sarah Hahn, Joe Engelberger, Blake Hannaford, Ruediger Dillmann, Matt Mason, Paul Verschure, Jean-Claude Latombe, Bernie Roth, Raoul Herzog, Shankar Sastry, Kimon Valvanis, James Crowley, Hal Varian, Paul Wright, Howard Moraff, Steve Antonson, Shawn Brixey, Dario Floreano, Catharine Clark, Karen Davidson, Steve Dietz, Rodney Douglas, Helen Grenier, Adele Goldberg, Réne Larsonneur, Caroline Jones, Jean-Daniel Nicoud, Philipp Bühler, Eduardo Kac, Stuart Dreyfus, Lev Manovich, Michael Naimark, Bruno Latour

Special thanks go to Katrin Schleiss and Marie-Jo Pellaud at EPFL for carefully correcting the text files and to our colleagues at UC Berkeley and the Swiss Federal Institute of Technology.

Introduction

Ken Goldberg and Roland Siegwart

Like many technologies, remote-controlled robots were first imagined in science fiction. They were put into practice in the 1940s for the remote handling of radioactive materials. Today, remote robots are used to explore undersea and deep in space, to defuse bombs and to clean up hazardous waste. Most of these robots are accessible only to trained and trusted experts. A new class of remote-controlled robots is now available to the general public that allow anyone with an Internet access anywhere to participate. This book describes eighteen such systems in detail.

Starting around 1993, a new Internet protocol, the HyperText Transfer Protocol (HTTP), facilitated the development of universal web browsers such as Mosaic, Netscape, Lynx, and Explorer. These browsers provide hyperlinked access to remote text and image files. Researchers quickly realized that in addition to prestored images, cameras could be set up to generate live images. By 2001, thousands of such *webcams* were accessible online [1].

Online robots go beyond webcams by including some mechanism for action. Although there is disagreement among experts about how to define *robot*, a robot is defined here as a physical device that is capable of moving through its environment, in contrast to a purely software-based "softbot" [4]. An online robot is a robot that is accessible from any computer on the Internet, generally through the HTTP protocol. Although the Internet provides a powerful means of access, it introduces a number of engineering challenges. Online robots take many forms and allow users to perform a variety of operations remotely. They must be carefully designed for nonspecialists and be engineered to run reliably around the clock.

Since the first online telerobot appeared in August 1994, about forty such systems have been put online by research teams around the world. This book summarizes the state of the art in online robot technology. It documents existing projects with sufficient engineering detail to allow readers to understand how these systems are designed, how they function, and the engineering challenges they solve. The book is intended for engineers, researchers, educators, students, and the general public. No prior knowledge of robotics is presumed.

For excellent reviews of telerobotics prior to the Internet, see [7] and [6]. Online robots have been the focus of two international research workshops [8, 9].

A related book, *The Robot in the Garden*, [2] addresses related social and philosophical issues.

This book includes eighteen chapters by eighteen research teams. Most chapters address one project in detail, describing its motivation, design of hardware and software, and performance statistics. Chapters are organized into four parts.

Part I introduces telerobotics and considers several remote manipulation systems. Craig Sayers provides a concise review of the basics of robot teleoperation. Sayers describes challenges common to all remote robots: time delay, limited sensory feedback, and user interface design. Jason, Robert Ballard's underwater telerobot used to explore the sunken Titanic, is used to illustrate these issues. Jason's robotic arm demonstrates the concepts of forward and inverse kinematics, redundancy, and singularities.

Sayers then defines teleoperation and shows how time delays complicate control. When an operator sees that the remote robot is moving and is about to smash into a rock, for example, she reacts by putting on the remote brakes, but the signal does not reach the remote robot in time to prevent collision. One solution, move-and-wait, is based on commanding very small motions and waiting for confirmation before sending the next command. This is obviously tedious for the operator. A better solution, teleprogramming, introduces some autonomous control at the robot, so that the robot can detect and respond before collisions occur. In such cases the remote operator is more of a supervisor than a puppeteer [7].

Chapter 2 describes the Mercury Project, the first telerobot to use the HTTP protocol and browser interface. A four-axis IBM robot with camera and air nozzle was set up over a sandbox so that remote viewers could excavate for buried objects by positioning a mouse and clicking from any web browser. Each operation is atomic (self-contained) and the application was designed so that singularities and collisions cannot occur. The system was designed to be operated by nonspecialists and to operate reliably twenty-four hours a day. Chapter 2 also describes the Telegarden and three other online robot projects that explored different interfaces and applications.

Chapter 3 by Barney Dalton gives a general framework for telerobotics and online robots. This framework grew out of the author's experience with Ken Taylor's Australian Arm, a six-axis robot manipulator and gripper that went online in September 1994, allowing users to build elaborate structures with colored blocks. Dalton's chapter addresses how state information can be maintained in the stateless HTTP protocol. As an alternative to passing tokens through common gateway interface (CGI) scripts, Dalton developed a Java-based client-server architecture that maintains state at the client to smoothly handle authentication and security.

Although the browser screen is inherently two dimensional, Hirukawa, Hara, and Hori describe 3-D interfaces based on VRML and Java in chapter 4. This teleprogramming approach simulates performance graphically prior to sending commands to the remote robot. They also propose a truly modular system where robots can be plugged into the network and controlled from any node using the CORBA protocol. As an application they consider a home robot controlled remotely that could check if the doors are locked and also load laundry and videotapes.

Part II emphasizes mobile robots that can be directed to move around in a remote environment. In chapter 5, Reid Simmons and his team at Carnegie Mellon University describe the first mobile Internet robot, Xavier. Online since 1995, Xavier accepts commands to visit classrooms and offices, broadcasting images (and even jokes!) as it goes. Local path planning and sensing is employed to avoid obstacles. Xavier has successfully traveled over 210 kilometers through hallways and doors in the Carnegie Mellon Computer Science Building.

In chapter 6, Saucy and Mondada describe Khep on the Web, a small mobile robot that moves within a maze. An elegant interface allows users to view either from the robot's perspective or from an external (bird's-eye) camera to choose driving directions. In addition to the physical setup, Saucy and Mondada provide a simulated 3-dimensional version of the maze and robot using Virtual Reality Markup Language (VRML). They report detailed statistics from over 27,000 visits in the course of a year.

Chapter 7, by Roland Siegwart and his collaborators, describes RobOnWeb, a system that includes five extremely small (2cc) mobile robots driven by Swiss watch motors. Each robot has an on-board camera capable of recognizing the other robots in the system. They move in a citylike scaled miniature environment and are controlled through a Java-based user interface that provides both graphical simulation and multiple camera perspectives simultaneously.

In chapter 8 Hashimoto, Ando, and Lee consider the human factors associated with perception of time delay on the Internet. They asked human subjects to drive a mobile robot through a maze under different time-delay conditions. They quantify how time delay affects navigation error and task completion times.

In chapter 9 Canny and Paulos consider the human interface implications of teleconferencing using online robots. They describer remote communication between humans and a blimplike robotic "space browser". In this case the robots are helium balloons driven by fans and controlled by a distant human operator who wishes to visit, say, a remote conference or science lab. Such floating or wheeled Personal Roving Presences (PRoPs) offer more engaging interaction and are equipped with cameras, microphones, and pointing devices to facilitate natural communication.

Part III focuses on issues related to control and time delay. Communication delays on the Internet can be lengthy and can vary significantly. As described earlier, the problems with time delays are particularly challenging on the Internet, where packets are routed dynamically and are not guaranteed to arrive in order. Variations in user load and time of day may lead to substantial variations in system performance.

In chapter 10 Brady and Tarn review major control paradigms used in teleoperation: predictive control, bilateral control, and teleprogramming. The latter describes a range of supervisory models where some degree of adaptive autonomous control occurs at the robot. The authors give a state-space formulation that treats variable time delays and events such as collisions, and they report on an experiment controlling a robot in St. Louis from Albuquerque, New Mexico.

Niemeyer and Slotine describe an alternative control architecture that ensures stability in chapter 11. They consider the problem of controlling a haptic (touch) feedback device over the Internet. Typical bandwidths for these devices when used locally are $100 - 500$kHz; on the Internet experiments show that time delays can reach 400 msec, which would cause standard feedback control to go unstable. Niemeyer and Slotine propose to use small packets that encode state information (velocity and force) using patented wave variables. This framework adapts gracefully to delays, causing the system to become sluggish rather than unstable when delays are substantial.

In chapter 12 Kosuge and colleagues suggest a different means to cope with variations in time delay. They extend the scattering transform method for fixed time delay using a quantity called virtual time delay, an estimate of the maximum expected time delay per feedback cycle. The authors incorporate a predictive display at the controller, which estimates the true position of the robot to help the human operator safely grasp and manipulate remote objects.

In chapter 13 Paul Backes and his team at JPL describe WITS, a web-based interface developed for NASA planetary vehicles on Mars. Since the time delay to Mars is particularly long and only available during bursts of a shared antenna, the mobile robot has substantial sensing and planning capacity built in. A web interface, including Java2 and Java3D, is used to provide a rich 3-D representation to humans to plan the next day's mission commands.

Hideki Hashimoto and Yasuharu Kunii describe two online robot systems in chapter 14. They report on an experiment controlling a mobile robot over the Internet; although it works fine in the mornings, data load is so high most afternoons that transmitted color images are sufficiently delayed to make control impossible. Hashimoto and Kunii also describe a pair of humanlike handshake devices that allow two remote humans to shake hands over the Internet. An impedance model with feed-forward control permits realistic performance with time delays up to 400 msec.

Part IV describes other novel applications from painting to chemistry. In chapter 15 Matthew Stein describes a Java-based teleprogramming framework developed for an online robot that allows users to paint with a brush. Users first download a custom Java applet that allows them to compose a painting using the four available colors. The resulting brushstrokes are then transmitted back to the robot for execution on paper (which is subsequently mailed to remote artists). Stein reports in detail on user behavior and gives examples of the resulting paintings.

In chapter 16 Burgard and Schulz address the problem of time delay for a mobile robot in a dynamically changing environment. In contrast to the teleprogramming model of chapter 4, Burgard and Schulz develop a probabilistic estimator of robot state that is displayed in a graphical 3-D simulation. Laser rangefinder data from the robot allow the position and velocity of objects to be updated using a maximum likelihood estimator and open loop path planner. These estimates provide efficient updates as the graphical simulator tracks robot state. They use UDP rather than TCP and demonstrate that their system performs far better than dead reckoning in the presence of time delays typical on the Internet.

In chapter 17 Steven Goldberg and George Bekey describe an online robot that allows both scholars and novices to examine a precious museum artifact, in this case a marble statue, remotely. The original artifact is placed on a rotating platform and viewed through a pair of stereo cameras mounted on a robot arm, all of which can be controlled remotely through a novel and intuitive interface.

One benefit of teleoperation is the ability to handle very small objects by spatially scaling human motions. In chapter 18 Venema and Hannaford employ scaling via the Internet to permit remote mounting of frozen protein crystals in glass tubes. Originally developed for microgravity experiments on the Space Station, their system coordinates a direct drive robot, linear micropositioners, and syringe pumps with several microscope video cameras to successfully perform these delicate operations via the Internet.

For the first time in history, telerobotics is widely available to the general public. This new hands-on potential has tremendous implications for education, allowing students to actively explore remote environments. It also has implications for research, where laboratories can share access to expensive resources such as coordinate measuring machines or scanning tunneling microscopes with other scientists. The Internet provides a standard interface but it also introduces new engineering challenges.

As we go to press, an American company has announced the first off-the-shelf online robot that comes with a mobile platform, extending neck, pan-tilt camera, and all necessary software (www.irobot.com). New applications for online robots are appearing every month. This book reflects the creativity, challenges, and excitement of this new field.

References

[1]   A. Ellin. You're on Constant Camera. *New York Times*. Sunday 19 Nov. 2000. p. ST2.

[2]   K. Goldberg. *The Robot in the Garden: Telerobotics and Telepistemology in the Age of the Internet*. Cambridge: MIT Press, 2000.

[3]   D. Kushner. Web Robots Offer Hands-On Experience From Afar. *New York Times*. Nov. 19, 1998. p. D4.

[4]   A. Leonard. *Bots: the Origin of a New Species*. Penguin Books, 1997.

[5]   mitpress.mit.edu/online-robots/,

[6]   C. Sayers. *Remote Control Robotics*. Springer Verlag, 1999.

[7]   T. Sheridan. *Telerobotics, Automation, and Human Supervisory Control*. Cambridge: MIT Press, 1992. See also Sheridan's "Musings on Telepresence and Virtual Presence." *Presence Journal* 1:1, 1992.

[8]   Organized by Roland Siegward. In *IEEE International Conference on Intelligent Robots and Systems (IROS): Workshop on Web Robots*. Victoria,

Canada, 1998.

[9]   Organized by Ken Goldberg and Eric Paulos. In *IEEE International Conference on Robotics and Automation (ICRA): Workshop on Current Challenges in Internet Telerobotics*. Detroit, MI, 1999.

Part I

Remote Manipulation

1 Fundamentals of Online Robots

Craig Sayers

1.1 Introduction

Online robots are constrained by the physics of operating a mechanical device, by the need to communicate via a low-bandwidth connection, by the unpredictability of a real-world environment, and by the need to interact with a human via a simple interface. Those constraints are not unique to the online world. Similar problems have been considered in traditional robotic systems — especially in a particular area of robotics termed teleoperation.

This chapter is aimed at readers who have an interest in online robots but who may be unfamiliar with the scope of prior relevant work in traditional robotics and teleoperation. It is not a sweeping review; instead, the focus here is on a few particularly interesting and relevant concepts, including references to more lengthy treatments.

The chapter starts by introducing basic robotics concepts using, as an example, the manipulator from a remote underwater vehicle. Then it introduces concepts from remote robotics, beginning with a robot connected via a dedicated wire link and then considering an interface using a computer workstation. Finally, it returns to a more web-centric focus, examining systems for controlling remote robots efficiently while using delayed, low-bandwidth, communications links.

In some sense, the problems faced by online robots are a superset of those faced in conventional teleoperation; and those, in turn, are a superset of those faced in conventional robotics. To make an online robot work requires an understanding of all three domains.

1.2 Robot Manipulators

This section examines those fundamental constraints that result from the physical design of the robot manipulator. For a more in-depth treatment, consult a good robotics text such as [10].

By way of example, the manipulator arm from the Woods Hole Oceanographic Institution's Jason underwater vehicle is used (figure 1.1). This manipulator is powered by six electric motors. The first five motors each control one joint on the

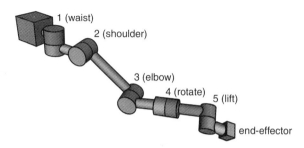

Figure 1.1:   *The manipulator arm from the Jason underwater vehicle provides an interesting example of a robot for operation under remote control. It has five joints, each powered by an electric motor.*

arm, while the last motor opens and closes the claw (or end effector).

The Jason arm has five joints (excluding the claw). In robot terminology, this means it has five degrees of freedom (DOF) where each DOF provides for motion about a different axis. The first three joints provide control over the position of the claw, while the last two joints control its orientation. In human terms, the first three joints act like a waist, shoulder, and elbow; while the final two act like a wrist, allowing the claw to rotate and lift. Controlling the Jason manipulator requires first determining where one wishes to move each joint. One approach is to control the robot in a relatively simple fashion by directly commanding desired joint angles. This is *joint space* motion.

An alternative approach is to take commanded motions in *Cartesian space* and map those into appropriate joint activations. The mapping from joint angles to the current end-effector location in Cartesian coordinates is called *forward kinematics* and, for manipulators like Jason, it is a simple matter of triginometry.

The reverse mapping, from a desired Cartesian location to joint positions, is called *inverse kinematics* and it is considerably more difficult. This is unfortunate, since it is precisely the computation one needs to permit a remote operator to command the manipulator using natural motions.

One complication is that a single Cartesian location may correspond to several joint positions. This is termed *redundancy*. For example, figure 1.2 shows two different joint positions with identical Cartesian end-effector locations.

The way to avoid redundancy is to introduce additional constraints. One could choose, for example, from among the possible solutions, the one that requires

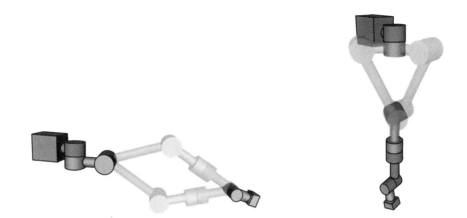

Figure 1.2:   Examples of a redundancy (left) and a singularity (right).

minimal joint motion, or minimal energy expenditure. Given some knowledge of what the operator was attempting, one could also make use of redundancy, for example, to maximize the force that the manipulator could exert along a particular axis [23].

Another complication is that the joint rotational axes may align in particularly inconvenient ways. Such a situation is shown in figure 1.2. Here joints one and four are aligned so that rotation of either has exactly the same effect. The result is that the benefit of having the fourth joint is lost while in that position. This is an example of a *singularity*. Software that converts an operator's desired Cartesian motion into joint positions must take the effect of redundancy and presence of singularities into account.

Moving the end-effector to an arbitrary Cartesian location and orientation requires at least six DOF (three for position and three for orientation). But the Jason manipulator has only five DOF. The result is that there are some positions/ orientations that simply cannot be reached. Such an example is seen in figure 1.3 which shows three possible orientations for a flask. In two of the three orientations there is a feasible joint solution, but for the third there is no solution.

If an operator were controlling the Jason arm remotely to pick up that flask, then some ingenuity would be required. One could push the flask into a more convenient orientation, pick it up by the ends instead of the middle, or move the

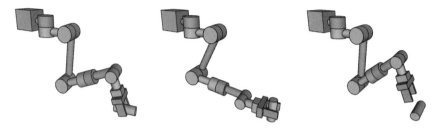

Figure 1.3:   Example of reaching for a flask. Since the arm has only five degrees of freedom, it is only able to grasp the middle of the flask in two of the three possible orientations.

base of the robot.

Even if the manipulator had six (or more) DOF, there will still be complications caused by the inevitable presence of joint limits and obstacles in the workspace. These can cause the Cartesian workspace to be discontinuous, and software must take these limits into account. One reasonable approach is to constrain the manipulator motion, preventing it from moving very close to a singularity or physical joint limit. If conditions allow, one can attempt to position the manipulator so common tasks may be performed without encountering motion limits and, if all else fails, provide the operator the option to fall back to direct joint-space control.

Computing where one wishes each joint to move is *kinematics*. The more difficult problem is making each joint move to where one computed. That requires an understanding of robot dynamics. In theory it might appear simple: one knows where each joint is, and one knows where one would like each joint to move. The complication is that physical factors intervene: inertia, friction, and external forces. Fortunately, for robots such as Jason, in which there is a high gear ratio between motor and actuator, and where the required accuracy is not excessive, these problems are quite tractable and solutions are reasonably well understood [10, 17, 23].

Given that one can take a joint-space, or Cartesian space, operator command and turn it into motion of the robot, one needs ways to let the operator enter that command and sense the resulting robot action. This general problem, of allowing an operator at one location, to control a robot at some other location, is called teleoperation.

1.3   Teleoperation

Teleoperation permits the performance of physical work at a remote site under operator control [21]. Typically, teleoperation systems consist of the following components:

- The operator.

- The operator station. In conventional teleoperation it would typically have a number of television monitors and a mechanical device with which one would command the system. In online robots it is often just a computer workstation.

- A remote robot. This is a mechanized contraption whose actions one wishes to direct from afar. It may be a manipulator like the Jason arm, or it may have legs, or it may even fly.

- Remote sensors. These may be as simple as manipulator joint position sensors, or as complex as television cameras. These serve two roles: they provide information for local control of the robot and they provide information to the distant human operator.

- Remote processor. This combines information transmitted from the operator station, with information from remote sensors, to generate control signals for the robot.

In the case of the Jason manipulator, the operator uses a hand-controller to command the robot and cameras to view the result on live television monitors. Position sensors are located in each joint and their information is combined with the operator's commanded motion in the robot controller. In this example, remote processing is performed using an array of transputers in a titanium pressure housing.

Communication between operator station and that remote processor is via ASCII characters sent down a serial link. That serial link is bidirectional and also provides information back to the operator display to control status lights and warn of problems. Further communication is via several live video streams from cameras at the remote site to monitors in front of the operator. All those signals, both serial and video, are transmitted via several kilometers of armored optical fiber cable. Typically, the operator station is located in a control van aboard a ship, while the Jason manipulator is mounted on the Jason vehicle and operated just above the sea floor.

When controlling the Jason manipulator, the operator may use the hand controller to control the manipulator in either joint or Cartesian space. In joint mode, each different motion of the controller maps to a different joint on the manipulator. In Cartesian mode, the motions of the controller generate Cartesian motions which are mapped into joint-space motions by the robot controller.

When using this system, the Cartesian control modes quickly become frustrating. One is forever bumping into joint limits and singularities. Often these are not obvious and the effect can be quite claustrophobic. Thus, operators quickly gravitate toward using joint-space control. This requires much practice. The mapping between controller motion and arm motion is particularly unnatural. Mastering it takes time and patience, but the results are surprisingly good. Experienced operators can move the arm with amazing dexterity.

Even experienced operators, however, still make occasional mistakes. The most common is to move a joint in the wrong direction. This might appear such an obvious error that it could be easily avoided, but its not so simple. For example, tilting the wrist up requires rotating joint five. But the direction depends both on the current arm configuration (especially how far joint four has been rotated) and on the relative position of the camera through which the operator is viewing the manipulator. Even the most experienced operators can often be seen making a small exploratory motion first, just to make sure they have the correct direction.

Jason's teleoperation system is deliberately simple. It is designed to work continuously for weeks at a time, and must do so, while several days sailing time from the nearest hardware store or computer programmer. When performing teleoperation in more congenial settings, or for more repetitive tasks, it is worthwhile to expend some effort to increase the efficiency with which operators may work.

When operating the Jason arm, one has the luxury of a direct wired link. The delay is imperceptible, and it is reasonable to send several live video streams. As one moves to teleoperation via the Internet, the delays increase and available bandwidths decrease. Thus one needs ways to compensate for those changes. Fortunately, there's already been a considerable amount of work done on teleoperation via constrained communications. The following sections explore some of these ideas.

1.4   Teleoperation on a Local Network

For this discussion, the direct teleoperation interface to the remote robot is replaced with an interface via a computer workstation connected via a local network. Assume here that the network has no perceptible delay.

The interface is now a computer and mouse, rather than television monitors and hand controller. To implement such a system, one needs to digitize the imagery from the remote site, send it over the network, and display it on the monitor. Then one needs to provide an interface through which the operator may command the remote manipulator.

To begin, a simple display of the live video image is provided in a window for the operator with sets of on-screen sliders for controlling the position of the manipulator in both joint and Cartesian space. Such a system may work surprisingly well, its simple to comprehend, and operators are quite adaptable. Nevertheless there is still much room for improvement.

To give the operator a better view of the remote site, one could add additional remote cameras and place cameras on mechanized platforms. Then one could automate the motion of those cameras and selection of imagery [3, 19, 22].

To provide additional information to the operator, one could calibrate the remote cameras [20] and then, using that knowledge, overlay visual clues onto the imagery to aid the operator in judging distances and alignments [8].

Given the luxury of a model of the remote environment, one could overlay that for the operator as well. Providing the operator with a virtual view that covered a wider field-of-view, and showed more detail than would be available from real camera imagery alone [12].

One of the problems of working remotely is the need for the operator to constantly map the desired end-effector motion into sequences of joint or Cartesian motions. A desired action "move a little to the left in this picture", for example, could easily turn into several commands if the "left in this picture" happened not to conveniently lie along a predefined Cartesian or joint motion axis. The solution is to remap the operator's commanded motions based on the viewing direction (figure1.4). In robotics terminology, this is considered an operator-centered frame of reference [13]. In computer graphics, it is termed *kinesthetic correspondence* [2]. The work required to map the operator's viewer-centric motions into

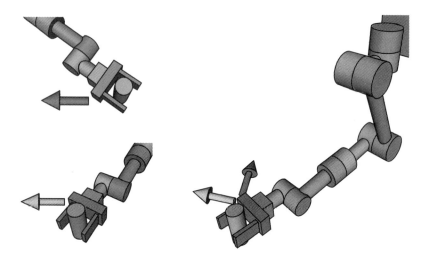

Figure 1.4: Providing a natural operator interface requires mapping operators' commanded motions based on their view of the remote site. For example, here is the same desired operator command, "move left", shown in two different views (left) along with the resulting desired manipulator motion (right).

manipulator commands may be minimized by carefully locating the cameras at the remote site. Fixed cameras may be aligned with the Cartesian base frame for the manipulator. Smaller cameras may be mounted to the end-effector and aligned with the end-effector frame of reference. The best solution, and the one that gives complete freedom in camera placement, is to calibrate the remote cameras and perform the required transformation math.

1.5   Teleoperation via a Constrained Link

Now, consider a communications scheme that imparts a several-second round-trip delay and limits the available bandwidth. One must be concerned with the effects of that constrained channel and perceptible delay.

The low bandwidth means one can no longer simply digitize and transmit imagery from the remote site as before. One simple technique is to choose a small image size and compress each image (for example, using JPEG compression). In this case, processing time is traded off (to encode and decode each image) in return for lower transmission bandwidths.

To cope with the delay, one convenient approach would be to do nothing and rely on the operator to adapt. Although attractively simple, such an approach will not work well. Consider the case where the operator notices the remote manipulator is about to hit a wall. The operator reacts immediately, but, by the time the image of the robot is seen, it is already several seconds too late. The collision happens long before the operator's desperate attempts at recovery make their way to the distant site. Knowing that risk, operators quickly adopt a "move and wait" strategy: making a small, relatively safe motion, then waiting to see that it succeeded before moving again [5]. This is frustratingly slow.

Coping with the delay requires a two-pronged approach. First, at the remote site, one must add some intelligence to the robot controller [9, 16, 18]. One does not need a very smart robot, but one needs enough intelligence to imbue it with a sense of self-preservation and the ability to protect itself and its environment. The controller need not know what has gone wrong or how to fix it; it has to react fast enough and protect itself and the environment for long enough so that the distant human operator has time to react. By providing the slave robot with additional information a priori, one can mitigate the effects of the delay.

By making the slave robot "smarter," one can also raise the level of communication between operator and remote robot, enabling the operator to assume a more supervisory role [16]. Even if the communications were not delayed, this is still beneficial. It frees the operator from the need to command every detail of the operation at the remote site.

The second approach to dealing with a constrained communications link is to supply the operator with a more sophisticated interface [6, 7, 15, 16]. For example, one could give the operator a simulation of the remote robot and allow for interaction with it. That's called a *predictive display* [1]. It allows the operator to see the effect of commands well before they are actually executed by the remote robot. If the remote environment were sufficiently constrained, then one could even consider simulating the dynamic motion of objects at the remote site [7]. By providing the operator station with a model of the remote robot that reacts instantaneously, one can largely insulate the operator from the effects of the communications delay.

Even if the link delay were not an issue, there may still be some benefit in providing the operator with a simulated view of the remote site. Such a system affords viewpoints unavailable from real cameras, it is unaffected by poor visibility at the remote site, and it frees the operator from the need to work at exactly the same speed as the remote site. The operator may experiment with actions off-line, or work more quickly or more slowly, than the real remote robot [4].

A number of working systems have been constructed for performing teleoperation via a constrained communications link. See, for example, the work of Sheridan et al. on supervisory control [16], Paul et al. on teleprogramming [11], and Hirzinger et al. on the ROTEX project [7].

All these approaches have a common theme: by providing additional information at each site, one can reduce the effects of delayed communications. The difficulty is that the bandwidth is also constrained. Thus, one must make trade-offs in sending information between sites. Having additional information may save time, but sending that information itself consumes time. Deciding which information to send between sites, and when to send it, are perhaps the most difficult and challenging part of efficiently controlling a remote robot via a constrained link [15].

1.5.1   Handling the Unexpected

Now, a predictive display exists for the operator. One can send commanded motions to a remote robot and, using knowledge of robot kinematics and dynamics, one can make the robot move to where it was commanded. It might appear that one has a complete system, but unfortunately that is less than half-way. What has been neglected is that the real world is far from perfect. Regardless of how exacting computer prediction might become, it will never be able to anticipate everything that the real world contains. Thus, merely sending commands to the remote robot is not sufficient. One must have ways to detect, and recover from, unexpected events.

In the teleprogramming system, for example, errors are detected by comparing predicted sensor readings (encoded within commands from the operator station) with actual sensor reading (measured as the slave executes each command). Any significant discrepancy causes the slave to pause and an error to be sent to the operator station.

Each error is expensive. When the error occurs, a signal has to propagate from the remote site back to the operator, and then the operator's new posterror commands have to travel to the remote site. Thus one loses at least one round-trip communications delay. In practice, the cost of the error is much greater. It takes time for the operator to diagnose each problem. Even worse, an error will often result in a discrepancy between the operator station model of the remote site and the real remote site. Regardless of how perfect remote site sensors might become, resolving discrepancies in the world model will always consume valuable time and bandwidth resources. Because of these costs, it is worthwhile to expend some effort to try and avoid errors or at least mitigate their effects [15].

While one can not predict and avoid everything that might go wrong, one can at least try to make sure that the correct commands are sent to the remote robot. This requires the same good basic user-interface design as in a conventional teleoperation system. In addition, one should aid the operator in performing actions that are likely to be successful, while dissuading the operator from actions that are likely to fail [15].

In addition to expending effort up-front to avoid errors, one can also expend resources to mitigate the cost of possible errors. For example, one can periodically send joint and other sensor information from the remote site back to the operator station. Then, if an error occurs, much of the information necessary to replay recent events will be available for the operator to aid in diagnosis.

Since there is insufficient bandwidth to send back all sensory information collected at the remote site, one must be selective. Due to the communications delay and the desire not to overburden the operator, one must perform that selection autonomously. The solution is to use knowledge of the remote site, the operator's commanded actions, and the relative likelihood of errors, to predict that fragment of the available sensory information that would best assist the operator if an error were to occur. This process is termed *intelligent fragmentation* [14].

With intelligent fragmentation, one makes the sensors at the remote site responsive to motion of the remote manipulator. The reverse is also possible. That is, one can make the motions of the manipulator dependent on an ability to sense it. While there is often little freedom in the manipulator positions (they are largely dictated by the operator's desired commands) one does have control over the speed of motion. In cases where errors are likely, one can automatically slow the speed of arm motion, allowing the sensors to capture each nuance.

Even if it were possible to send a large volume of sensory data to the operator station, there will still be a need to select some fragment for display. Ultimately, it will be the operator's bandwidth, and not the communications bandwidth, that is the final limiting factor.

1.6   Conclusions

To cope with the low bandwidth and high latency imposed on current online robots, one needs to make trade-offs when sending information between sites. The benefit of having additional information must continually be weighed against the cost of getting it to the other site.

Online robot systems will draw upon ideas from conventional teleoperation, especially teleoperation via constrained communications. Expect images from remote cameras to be compressed intelligently, taking into account information about the task and scene rather than just raw pixel values. Expect interfaces to become more abstract, and for the level of interaction between operator and machine to rise. This will have the dual benefit of simplifying the operator interface and reducing the bandwidth requirements for signals sent from master to remote site.

The most difficult part of constructing workable online robots, particularly those for use outside carefully engineered environments, will be dealing with all those things that might go wrong. Its not necessary to predict each possible failure in advance, but it is necessary to provide the operator with the means to diagnose, and then recover from, any unexpected event. Expect much more intelligent remote robots and much more sophisticated operator interfaces. These will allow the remote robot to detect problems promptly and allow the operator to diagnose and correct them quickly.

The Internet will certainly improve. As the delays decrease and bandwidths increase, online robots will become much more like conventional teleoperation systems. The unique characteristics of online robots will then be dictated by their intended audience and the limited operator interface available via a computer screen.

Interestingly, many of the concepts from this chapter will still be applicable then. The need to interact efficiently with a 2-D image will still be important, many of the advantages of a simulation or predictive display will still be present, and the automatic selection of some fragment from the available sensory information will still be necessary.

Having an intelligent human "in the loop" is a powerful ally. Designs should strive to make the best use of that intelligence, with the operator solving the difficult problems, while the system takes care of the details.

Acknowledgments

Much of the author's knowledge of robotics is the direct result of many thoughtful conversations with his thesis adviser, Richard Paul. Those discussions indirectly influenced this work.

The Jason subsea vehicle and manipulator system was designed and developed at the Woods Hole Oceanographic Institution (WHOI). A number of people contributed to it, including Bob Ballard, Andy Bowen, Tom Crook, Dick Edwards, Bob Elder, Larry Flick, Skip Gleason, Bevan Grant, Bob McCabe, Tad Snow, Will Sellers, Cindy Sullivan, Nathan Ulrich, Louis Whitcomb, and Dana Yoerger. It is operated by the WHOI Deep Submergence Operations Group, supported by NSF Grant OCE-9300037. The author's experience in using it for remote robotics was funded in part by the National Science Foundation and the Naval Underwater Warfare Center. The author is especially grateful to the Woods Hole Oceanographic Institution for providing a postdoctoral scholarship.

References

[1]   Anatal K. Bejczy, Steven Venema, and Won S. Kim. Role of computer graphics in space telerobotics: Preview and   predictive displays. In *Cooperative Intelligent Robotics in Space*, pp. 365–377. Proceedings of the SPIE, vol.1387, November 1990.

[2]   Edward G. Britton, James G. Lipscomb, and Michael E. Pique. Making nested rotations convenient for the user. In *ACM SIGRAPH*, pp. 222–227, 1978.

[3]   Thurston Brook,s Ilhan Ince, and Greg Lee. Vision issues for space teleoperation assembly and servicing. *STX Robotics Corp. Technical report STX/ROB/91-01*, Maryland, 1991.

[4]   Lynn Conway, Richard Volz and Michael Walker. Tele-autonomous systems:Projecting and Coordinating Intelligent Action at a Distance. *IEEE Robotics and Automation Journal*,vol. 6, no. 2, 1990.

[5]   William R. Ferrell and Thomas B. Sheridan. Supervisory control of remote manipulation. *IEEE Spectrum* 4 (10): 81–88, October 1967.

[6]   Janez   Funda.   Teleprogramming–toward   delay-invariant   remote manipulation. *Ph.D. dissertation*. The University of Pennsylvania, Philadelphia, 1992.

[7]   G. Hirzinger. ROTEX the first robot in space. In *International Conference on Advanced Robotics*, pp. 9–33,   November 1993.

[8]   Won S. Kim and Lawrence W. Stark. Cooperative control of visual displays for telemanipulation. In *IEEE International Conference on Robotics and*

Automation. May, 1989.

[9]   Thomas S. Lindsay. Teleprogramming–remote site task execution. *Ph.D. dissertation.* The University of Pennsylvania, Philadelphia, 1992.

[10] Richard P. Paul. *Robot manipulators: mathematics, programming, and control.* Cambridge: MIT Press, 1981.

[11] Richard P. Paul, Janez Funda, and Thomas Lindsay. Teleprogramming: Toward delay-invariant remote manipulation, *Presence,*vol. 1, No. 1, 1992.

[12] D. Maddalena, W. Predin, and A. Terribile. Supervisory control telerobotics reaches the underwater work site. *Proceedings of the 6th International Advanced Robotics Program Proceedings on Underwater Robotics.* Toulon, France, 1996.

[13] Daryl Rasmussen. A natural visual interface for precision telerobot control. In *SPIE Telemanipulator Technology*, vol. 1833, pp.170–179,   Boston, MA, November 1992.

[14] Craig Sayers. Intelligent image fragmentation for teleoperation over delayed low-bandwidth links. In Proc. *IEEE International Conference on Robotics and Automation.* May 1996.

[15] Craig Sayers. *Remote Control Robotics.* New York: Springer Verlag, 1998.

[16] Thomas Sheridan. *Telerobotics, Automation, and Human Supervisory Control.* Cambridge: MIT Press, 1992.

[17] Mark Spong and M. Vidyasagar. *Robot Dynamics and Control.* Wiley, 1989.

[18] Matthew R. Stein. Behavior-Based Control For Time Delayed Teleoperation. *Ph.D. thesis*, University of Pennsylvania MS-CIS-94-43.

[19] Konstantinos Tarabanis and Roger Y. Tsai. Sensor planning for robotic vision: A review. In Craig Khatib and Lozano-Perez, ed. *Robotics Review* 2, pp. 113–136. Cambridge: MIT Press, 1992. .

[20] R.Y. Tsai. A versatile camera calibration technique for high accuracy 3-D machine vision metrology using off-the-shelf tv cameras and lenses. *IBM Research Report* RC 51342, 1985.

[21] Jean Vertut and Philippe Coiffet. *Robot Technology*, vol. 3: *Teleoperations and Robotics.* Prentice-Hall, 1986.

[22] Yujin Wakita, Shigeoki Hirai, and Toshiyuki Kino. Automatic camera-work control for intelligent monitoring of telerobotic tasks. In *IEEE/RSJ International Conference on Robotics and Systems*,   pp. 1,130–1,135, July 1992.

[23] Tsuneo Yoshikawa. *Foundations of Robotics: Analysis and Control.* Cambridge: MIT Press, 1990.

2 The Mercury Project: A Feasibility Study for Online Robots

Ken Goldberg, Steve Gentner, Carl Sutter, Jeff Wiegley, and Bobak Farzin

2.1 Introduction

2.1.1 Motivation

Initiated at CERN in 1992, the HyperText Transfer Protocol (HTTP) provides a standard graphical interface to the Internet [1] that quickly attracted a worldwide community of users. In spring 1994, we considered how this technology could be used to provide offer public access to a robot.

As illustrated in figure 2.1, we set up a robot arm over a semi-annular workspace containing sand and buried artifacts. We attached a camera to the end of the robot arm along with a nozzle to direct air bursts into the sand. An interface was then developed so this hardware could be controlled via the Internet.

The primary criterion was that the system be reliable enough to operate 24 hours a day and survive user attempts at sabotage. A practical criterion was that the system be low in cost as there was a limited budget. It is worth noting that the manufacturing industry uses similar criteria, reliability and cost, to evaluate robots for production. Thus our experience with minimalist robotics [2] proved helpful.

A secondary goal was to create an evolving web site that would encourage repeat visits by users. Toward this end, all of the buried artifacts were derived from an unnamed 19th-century text. Users were asked to identify this text and thereby collectively solve the "puzzle". After each five-minute operating session, users were prompted to describe their findings and hypotheses in an ongoing operator's log. When the system was decommissioned the operator's log contained over a thousand pages of entries. The answer to the puzzle was announced in March 1995: Jules Verne's *Journey to the Center of the Earth*.

After discussing related work in the next section, we describe the system architecture, hardware, user interface, system performance, and subsequent online robot projects.

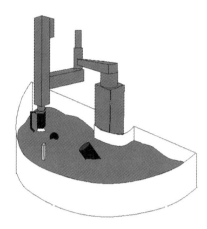

Figure 2.1:   Robot, camera, and air nozzle above workspace.

2.1.2   Related Work

Goertz demonstrated one of the first "master-slave" teleoperators fifty years ago at the Argonne National Laboratory [3]. Remotely operated mechanisms have long been desired for use in inhospitable environments such as radiation sites, undersea [4], and space exploration [5]. At General Electric, Mosher [6] developed a complex two-arm teleoperator with video cameras. Prosthetic hands were also applied to teleoperation [7]. More recently, teleoperation is being considered for medical diagnosis [8], manufacturing [9], and micromanipulation [10]. See Sheridan [11] for an excellent review of the extensive literature on teleoperation and telerobotics.

Most of these systems require fairly complex hardware at the human interface: exoskeleton master linkages are attached to the human arm to minimize the kinesthetic effects of distance to create a sense of "telepresence". The objective here was to provide widespread access by using only the interface available in standard HTML.

Internet interfaces to Coke machines were demonstrated in the early 1980s, well before the introduction of HTTP in 1992. The first webcam was the Trojan coffeepot at Cambridge University. Today there are thousands of webcams online [12].

Although the Mercury Project was the first online robot[1], it is important to acknowledge the independent work of Ken Taylor at the University of Western Australia. Taylor demonstrated an Internet interface to a six-axis robot with a fixed observing camera in September 1994 [13]. Although Taylor's initial system required users to type in spatial coordinates to specify relative arm movements, he and his colleagues have subsequently explored a variety of user interfaces [14].

In October 1994 Eduardo Kac and Ed Bennett exhibited a telerobotic artwork combining telephone and Internet control. Later that fall, Richard Wallace demonstrated a telerobotic camera and J. Cox created an online system that allows web users to remotely schedule photos from a robotic telescope [15]. Paulos and Canny have implemented several online robots with elegant user interfaces [16, 17]. Bekey and Steve Goldberg used a six-axis robot, rotating platform and stereo cameras to allow remote scholars to closely inspect a rare sculpture [18]. Since then there have been a number of Internet-controllable mobile robots [19,20,21] and NASA has used the Internet for control of vehicles in space [22]. Recently, Stein [23] demonstrated a Java interface to control a painting robot over the Internet.

Schulz [26], Slotine [27], and Tarn [28] discuss how control theory used for teleoperation over private lines can be adapted to the Internet, where delays vary unpredictably. They restrict access to their system to a single set of trusted users. The goal here was to develop a public teleoperated system that could be accessed by anyone on the Internet at any time. A variety of other online robots are reported in this book.

2.2   System Description

2.2.1   Hardware

The Mercury Project used the IBM SR5427 robot built by Sankyo in early 1980. Its four-axis design is common in industrial assembly for pick-and-place operations because it is fast, accurate, and has a large 2.5D workspace. This robot was selected over other robots in our lab due to its excellent durability, large workspace, and because it was gathering dust in a corner.

Unfortunately IBM no longer supported this robot, and the authors were forced

---

[1]   *Preliminary versions of this chapter appeared in the 1995 IEEE International Conference on Robotics and Automation and at the 1994 Second International WWW Conference.*

to read two antiquated BASIC programs and monitor their serial line transmissions to decipher the protocols needed for serial control of the robot. The robot accepts joint motion commands using IEEE format and check sums.

To allow users to manipulate the remote environment, the initial plan was to place a simple gripper at the end effector. Anticipating user attempts at sabotage (a time-honored hacker tradition), compressed air was used instead as the medium for manipulation.

The CCD camera was an EDC 1000 from Electrim Inc., chosen based on size and cost. Image data was sent from the camera back through a custom serial line to a video capture card. The camera image had a resolution of 192 by 165 pixels with 256 shades of gray, which were truncated to 64 shades to reduce transfer time. Exposure time could be changed by software to range between 64ms to 200ms. Although the robot was slowed to minimize dynamic effects, mechanical settling times were long enough to cause image blur at the camera. To avoid this, a stability check was implemented by taking two images separated by 64ms and differencing them. Subsequent images were then taken until the two successive images were sufficiently similar.

To avoid the complexity of another servo motor, a fixed focus camera was used and a focal point chosen that compromises between the two fixed camera heights. The workspace was primarily illuminated by standard florescent fixtures. A contrast enhancement routine was tested to normalize the lighting of each image captured from the camera. This increased image quality in most cases but exaggerated intensity variations across the workspace.

2.2.2   User Interface

To facilitate use by a wide audience of nonspecialists, all robot controls were made available via Mosaic, the only browser available at the time. Mosaic's point-and-click interface is shown in figure 2.2. This 2-D interface matched the workspace of the robot, so users could move by clicking on the screen with a few buttons for out-of-plane effects. Users were trained with an online tutorial prior to operating the robot.

The user interface centered around a bitmap that was called the "control panel" as shown in figure 2.3. Any number of observers could simultaneously view the control panel, but only the current operator could send commands by clicking on

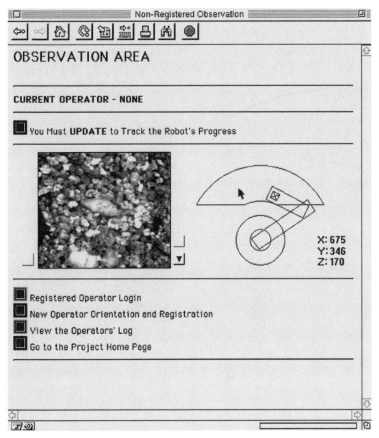

Figure 2.2:   The interface as viewed by a the Mosaic Internet browser.

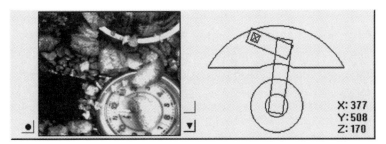

Figure 2.3:   The control panel. At the right is a schematic top view of the semiannular workspace
and robot linkage. At left is a camera image of the view directly beneath the robot end-
effector. Up/Down buttons are included for Z motion of the camera, and the round
button is used to blow a burst of compressed air into the sand.

Figure 2.4:   Sample camera images: top row shows scene before burst of compressed air, bottom row after. Left column taken by camera in the up position; right column taken by camera in the down position.

the image. To limit access to one operator at a time, a password authentication was implemented and a queue that gives each operator 5 minutes at the helm.

When the operator clicked on the control panel using the mouse, the XY coordinates were transferred back to the server, which interpreted them to decode the desired robot action. This action can be: (1) a global move to center the camera at XY in the schematic workspace, (2) a local move to center the camera at XY in the camera image, (3) moving the camera to one of two fixed Z heights, or (4) blowing a burst of compressed air into the sand directly below the camera (figures 2.4 and 2.5).

The size of the control panel was reduced to minimize turnaround time when a command is issued. The average image size for the control panel, encoded as a GIF file, was 17.3 Kbytes. Although response times of 10 seconds were achieved for on-campus users, cycle times of up to 60 seconds were reported from users in

Figure 2.5:   Composite image of workspace with artifacts such as miniature lantern, seed packet, and so forth.

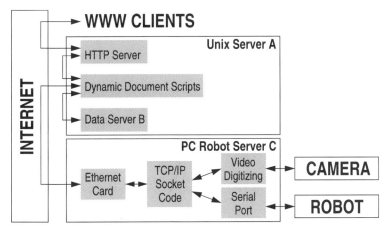

Figure 2.6:   System architecture,

Europe operating via 14.4K telephone lines.

A fictional context was also created for the system, inventing the history of a deceased paleohydrologist who had discovered unexplained artifacts in a radioactive region of southwest Nevada. It was noted that the Mercury robot was originally developed to explore the region and one mandate of the grant was to make the system "available to the broader scientific community." A hypertext document describing this background provided an online introduction.

2.2.3   Software Architecture

As shown in figure 2.6, users (web clients) accessed the system through the

Internet. The system included three communicating subsystems. Server A ran the vanilla NCSA HTTP Demon v.1.3 on a Sun SPARCserver 1000, with SunOS release 5.3. Server A cached the most recent control panel and sent it whenever an observer request came in.

When a user registered as an operator by entering a password, a database server was used to verify. This custom database server, B, ran on the same machine as server A.

After an operator was verified, server A either added the operator to the queue or communicated with server C which controls the robot. Server A decoded the X and Y mouse coordinates and sent them   to server C via the Ethernet.

On server C, a custom program decoded the XY coordinates into a robot command and verified that the command was legal, that is within the robot workspace. If so, the command was then executed via a command sent to the robot over a 4800-baud serial line. Once the command was completed, server C used a local frame buffer to capture the image.

Server C was a PC running MS-DOS (there was no camera driver for Linux at that time). Server C generated a new schematic view of the robot in the resulting configuration, then combined it with the camera image and appropriately highlighted control buttons to form a new control panel. Server C then compressed this image into GIF format and returned it to server A, which updated the most recent control panel and returned it to the client.

To maintain compatibility with the widest possible set of user platforms, only the HTTP protocol that was standard in 1994 (version 1.0) was used, thus restricting the project to single still GIF images; today streaming video is standard on most browsers.

The major difficulty in implementing server C was scheduling responses to the network, the local mouse, and the image capture board. Although a multitasking environment such as Linux was preferred, the Electrim camera was only compatible with DOS and the company would not part with any source code. Thus hand-crafted memory management was utilized and the screen itself was used as a memory buffer. This enabled a custom GIF encoder to be speeded down to a few microseconds per control panel.

2.2.4   Discussion

It turned out that a fundamental problem was how to handle multiple simultaneous viewers, delivering the appropriate web page to the site that requested it. To accomplish this we developed a system of random tokens.

Random Tokens

Each time server A returned a new control panel to an operator or observer, it added a large random number to its embedded URL for the update button. This random token prevented the client from caching the control panel (otherwise repeated requests to update the image would simply reload the local image and not request an updated image from server A).

The random token also allowed server A to identify and track clients. When an operator logged in with a verified password, server A tracked the operator by maintaining a database of recent clients so that URL requests could be customized depending on the user's status. For example, the queue was only visible to the operator and those on deck.

Servers A and B were at opposite ends of the USC campus and were connected via Ethernet. Each machine had its own IP address and resided in the usc.edu domain. Communication was achieved using a socket connection between the two machines. The implementation on server A was done using the standard BSD socket functions provided with the SunOS 4.1 operating system and Perl. On server C, the PC, a publicly available socket package called Waterloo TCP and Borland C were used.

2.3   Performance

2.3.1   Dates of Operation

The Mercury Project was online for seven months starting in late August 1994. During this time it received over 2.5 million hits. An off-center air nozzle caused sand to accumulate in one side of the box, so the sand had to be periodically regroomed. Other than a few random downtimes, the robot was online twenty-four hours a day.

Traffic Statistics

System usage with was monitored standard access logs and with custom logs at server B. If a "hit" is defined as a client request for a file, 1,968,637 hits were made by 52,153 unique hosts (figure 2.7). If "uses" were defined as clusters of hits with less than half hour idle time, the system was used 87,700 times due to repeat

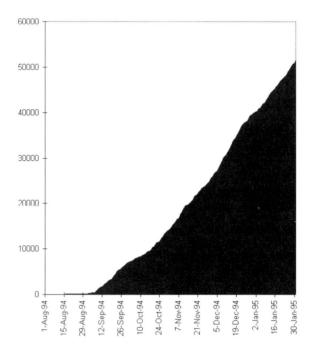

Figure 2.7:   Cumulative number of unique (new) hosts accessing the project.

visits. The daily average was 430 uses generating approximately 1,000 new images. In 1994 the Mercury Project accounted for roughly half of all requests to USC's web server during that period.

Network throughput averaged 20 Kbytes/sec, which was poor compared with 500 Kbytes/sec that can be achieved between two workstations in close proximity on the campus network. This was blamed largely on the MS-DOS operating system which forced the implementation of busy/wait cycles to obtain concurrence between the robotic/camera operations and the networking duties.

When server C detected an error, it automatically reset the robot controller, recalibrated, and returned the robot to its previous position. Also, server A automatically sent out e-mail if any of the key servers stopped functioning. This occurred on average twice a month usually due to restarts of the primary USC server. Server A also sent mail to the project team when server C stopped responding, which occurred about once a month.

2.3.2   Discussion

Operator Logs

Some samples from over 1,000 pages of operator's logs:

From: Rex Kwok
Date: Thu Nov 3 21:52:17 PST 1994
"It is amazing to operate a robot arm from Australia."

From: Scott Hankin
Date: Fri Sep 23 09:34:59 PDT 1994:
"...this site seems similar to the Internet. The search
is analogous to trying to find something on the net,
where you scan to locate areas of interest. Sometimes
you'll encounter a useful nugget of information like
[the antique lantern]; other times you'll discover
information which seems valid but may be misleading,
like the sample of "fool's gold". Some information is in
different languages, like the scrap of paper with text
in English and German which points to the multinational
nature of the net."

From: Dr. Steve M. Potter
Date: Thu Oct 27 23:30:09 PDT 1994
"What fun! Cool idea, folks. Best use of forms and clickable maps I have seen...I was wondering how I know this
is not a clever laser-disk full of pictures you grabbed,
with no robot, until i saw the time on the watch after
blasting it. That was when my skepticism evaporated."

From: James Bryant
Date: Sat Sep 10 08:54:11 PDT 1994
"I don't believe I have seen a nicer application of
science, or its money on the net."

Figure 2.8:   The Telegarden.

The Mercury Project was an example of minimalist robotics [2]. The robot had only four axes, is relatively inexpensive and robust, and it was easy to avoid singularities.

The Mercury Project was a feasibility study for a broad range of applications using the Internet to bring remote classrooms and scientific lab equipment to a much wider audience. Since the HTTP protocol offers no transmission time guarantees, this interface design is not suitable for time-critical interactions such as remote assembly with force feedback. The Mercury Project required operators to work sequentially and wait in a queue. The project was followed with a second online robot project, the Telegarden.

2.3.3   The Telegarden

The Telegarden [29] allows users to view and interact with a remote garden filled with living plants (figure 2.8). The Telegarden has been online since in August 1995. This project used a faster Adept-1 robot and a multitasking control structure so that many operators could be accommodated simultaneously: *http://telegarden.aec.at/*

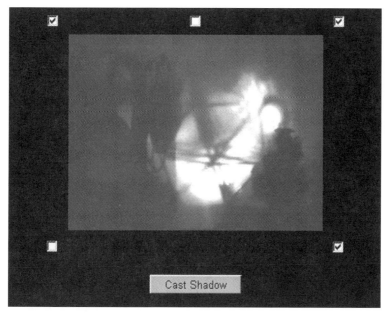

Figure 2.9:   A shadow image.

2.3.4   Dislocation of Intimacy

Our third interactive Internet system was Dislocation of Intimacy [30]: *http:// www.dislocation.net*

The goal in this project was to create a simple, fast, reliable, relocatable telerobotic system. This system was intended to be accessible to anyone with a browser. The project used lighting illumination as its action and returned live images to the user to show the results of that lighting combination at that instant. The system needed to function unattended at all times and cope with low lighting conditions.

The user was presented with a simple user interface that allowed remote control of the lighting conditions within the light box. The first page presented to the user gives a description of the project and presents the user with a "proceed" button. The "proceed" button invokes the server application and returns a web page, as seen in figure 2.9, with one randomly chosen light in the on position. Lighting control is indicated with the check boxes, where a checked box indicates

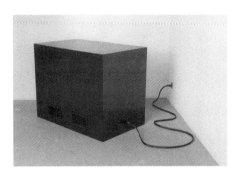

Figure 2.10: Lightbox exterior.

an illuminated light (top center.) The client is then able to activate or deactivate any of the check boxes and then by pressing the "cast shadow" button submit the request to the server.

The server processes the request and returns the appropriate shadow image to the client with the unique image from the lightbox and the check boxes marked according to the request. The client can then select any other combination of lights and repeat the process by clicking on the "cast shadow" button again. There is no limit to the number of requests a user can make. Although there are five buttons with a possible thirty-two lighting combinations, because of random motion in the box, there is no limit to the number of shadows that can be formed.

On each request page, there was a link to a document that allows the user to e-mail feedback, read the "most promising inferences," and review the most current statistics of the site.

System Architecture

Computer Hardware

The lightbox (see figure 2.10) sits 58 inches deep, 38 inches wide, and 48 inches high. The Pentium PC computer systems ran Red Hat Linux 5.0. Linux offers several advantages over Windows NT as a development environment for telerobotics. The Linux Apache 1.3.0 web server is more stable than comparable NT systems, and Linux allows greater access to the system memory and devices and is easily adaptable to   custom TCP/IP software and controller hardware.

Figure 2.11: Lightbox interior.

Physical Apparatus

The black steel exterior has one cable connecting it to the outside world; it was provided with a cooling and ventilation system. Figure 2.11 illustrates the interior of the apparatus box. At the near end is the control box containing the cooling system that moves air over the power transformers operating the lights and one of the computer systems. The air is directed out the lightbox through four vents that surround the exterior.

In the middle of the box is a scrim made of Acrylite FF, a thin semitransparent acrylic that serves as a screen for the projected shadows. The original screen was paper, but to avert fire hazard a switch was made to the semitransparent acrylic panel that is inexpensive, durable, and of the proper opacity to allow a crisp shadow.

At the far end are the lights that illuminate the objects, which are placed between the lights and the scrim. The lights are switched from the control box with custom circuitry that toggles a set of six solid state relays.

Semaphore

Because of the rapid processing of requests and the lack of a queue in the system, a semaphore was implemented that alternately creates and destroys a lock file.

The semaphore allows the execution of a critical region of code that requires exclusive use of the lightbox. The first efforts attempted to use the UNIX procmail v3.10 lockfile() command. This command however forks a child process, establishes a new shell, and then writes out the lock file. This shell is kept alive until the critical region is completed and the lock is released. Although this system appear to work well under a light load, when upward of 1,000 hits a day occurred, the server would crash because the spawning and killing of shells was OS intensive and takes up to 1 second, where the file manipulation within C takes 2 to 5 milliseconds.

Instead, the process simply checks for a lock file to become available and does not use the Unix command. When a request arrives, application checks for the presence of the "lock" file; if present, it waits for a random 0 to 500ms, and then tries again to access the "lock" file. This random waiting time avoids deadlock. Because this is all coded into the application, only a single application is open at any one time, and there are no child processes. Also, the maximum wait time between checking for a free lock is 500ms, which cuts the wait time and number of processes on the server in half, more than doubling the overall speed of the system.

Performance

From 21 April 1997 to 3 August 1998, there were 350,000 requests from 30,000 unique hosts. The server averaged 180 requests per day, about 8 per hour, but received two major spikes in usage resulting from press releases on WiReD news coupled with selection as a Yahoo! Pick of the week and a mention in the New York Times online. Over this time period, the average user requested three shadow images per visit, and the majority of the traffic was referred from news stories and search engine listings.

The system then went off line for several days while it was moved to an exhibition in San Francisco. The project then went back on line. From 3 August 1998 to 26 January 1999 it received 45,000 hits, from over 8,000 unique hosts. Over the course of the project we received over 70 e-mail responses. Ten messages of note can be found at: http://taylor.ieor.berkeley.edu/shadowserver/topten.html

2.3.5   Ouija 2000

Our fourth project focuses on collaborative teleoperation. A system was built that allows a distributed user group to simultaneously teleoperate an industrial robot arm via the Internet [31]. A Java applet at each client streams mouse motion vectors from up to twenty simultaneous users; a server aggregates these inputs to produce a single control stream for the robot (figures 2.12 and 2.13). Users receive visual feedback from a digital camera mounted above the robot arm. To the authors' knowledge, this is the first collaboratively controlled robot on the Internet. To test it please visit: http://ouija.berkeley.edu

Figure 2.12:   The Ouija 2000 user interface.

Figure 2.13:   The Robot with pointer.

Acknowledgments

Michael Mascha and Nick Rothenberg from USC's Department of Anthropology played an active role in the design of the Mercury Project. Also providing support and helpful suggestions were Howard Moraff , John Canny, Eric Paulos, Richard Wallace, George Bekey, Andy Fagg, Juergen Rossman of the University of Dortmund, Peter Danzig, Rick Lacy, Mark Brown, Eric Mankin, Irene Firtig, Victoria Vesna, Peter Lunenfeld, the Los Angeles Museum of Miniatures and the Laika Foundation. The Telegarden project was co-directed by Ken Goldberg and Joseph Santarromana (UC Irvine). The project team included George Bekey,

Steven Gentner, Rosemary Morris, Carl Sutter, and Jeff Wiegley. The director would also like to thank Erich Berger, Elizabeth Daley (USC-ACC), Len Silverman (USC-Engineering), Peggy McLaughlin (USC-Annenberg), Eric Mankin (USC News Service), and Rick Wagner. Dislocation of Intimacy was codirected by Goldberg and Farzin. We would also like to thank Adam Jacobs, Eric Lee, Leah Farzin, Eric Paulos, Derek Poon, Andrew Ludkey, and Tiffany Shlain. Ouija 2000 was directed by Ken Goldberg. The project team includes Rory Solomon, Billy Chen (Java Design/Coding), Gordon Smith (Control Design), Jacob Heitler (Java Design), Steve Bui (Camera), Bob Farzin (Hardware Design), Derek Poon (Robot Interface), and Gil Gershoni (Graphic Design). Thanks also to everyone who participated online.

References

[1] Tim Berners-Lee, Robert Cailliau, Jean-Francios Groff, and Bernd Pollerman. World-wide web: The information universe. *Electronic Networking:Research, Applications and Policy,* 1(2), Westport, CT, spring 1992.

[2] John Canny and Ken Goldberg. A RISC approach to sensing and manipulation. *Journal of Robotic Systems,* 12 (6): 351-362, June 1995.

[3] Raymond Goertz and R. Thompson. Electronically controlled manipulator. *Nucleonics,* 1954.

[4] R. D. Ballard. A last long look at Titanic. *National Geographic* 170(6), December 1986.

[5] A. K. Bejczy. Sensors, controls, and man-machine interface for advanced teleoperation. *Science* 208(4450), 1980.

[6] R. S. Mosher. Industrial manipulators. *Scientific American* 211(4), 1964.

[7] R. Tomovic. On man-machine control. *Automatica* 5, 1969.

[8] A. Bejczy, G. Bekey, R. Taylor, and S. Rovetta. A research methodology for tele-surgery with time delays. In *First International Symposium on Medical Robotics and Computer Assisted Surgery*, September 1994.

[9] Matthew Gertz, David Stewart, and Pradeep Khosla. A human-machine interface for distributed virtual laboratories. *IEEE Robotics and Automation Magazine*, December 1994.

[10] T. Sato, J. Ichikawa, M. Mitsuishi, and Y. Hatamura. A new micro-teleoperation system employing a hand-held force feedback pencil. In *IEEE International Conference on Robotics and Automation (ICRA).* May 1994.

[11] Thomas B. Sheridan. *Telerobotics, Automation, and Human Supervisory Control.* Cambridge:MIT Press, 1992.

[12] Search yahoo.com under `InterestingDevices'.

[13] Barney Dalton and Ken Taylor. A framework for internet robotics. In *IEEE International Conference On Intelligent Robots and Systems (IROS): Workshop on Web Robots*. Victoria, Canada, 1998.

[14] http://telerobot.mech.uwa.edu.au/

[15] http://www.eia.brad.ac.uk/rti/

[16] Eric Paulos and John Canny. Delivering real reality to the world wide web via telerobotics. In *IEEE International Conference on Robotics and Automation (ICRA)*. 1996.

[17] E. Paulos and J. Canny. Designing personal tele-embodiment. In *IEEE International Conference on Robotics and Automation (ICRA)* Detroit, MI, 1999.

[18] Yuichiro Akatsuka, Steven B. Goldberg, and George A. Bekey. Digimuse: An interactive telerobotic system for viewing of three-dimensional art objects. In *IEEE International Conference on Intelligent Robots and Systems (IROS): Workshop on Web Robots*. Victoria, Canada, 1998.

[19] Reid Simmons. Xavier: An autonomous moboile robot on the web. In *IEEE International Conference on Intelligent Robots and Systems (IROS): Workshop on Web Robots*. Victoria, Canada, 1998.

[20] Patrick Saucy and Francesco Mondada. Khep-on-the-web: One year of access to a mobile robot on the internet. In *IEEE International Conference on Intelligent Robots and Systems (IROS): Workshop on Web Robots*. Victoria, Canada, 1998.

[21] Roland Siegward, Cedric Wannaz, Patrick Garcia, and Remy Blank. Guiding mobile robots through the web. In *IEEE International Conference on Intelligent Robots and Systems (IROS): Workshop on Web Robots*. Victoria, Canada, 1998.

[22] Kam S. Tso, Paul G. Backes, and Gregory K. Tharp. Mars pathfindes mission internet-based operations using wits. In *IEEE International Conference on Robotics and Automation (ICRA)*. Leuven, Belgium, 1998.

[23] Matthew R. Stein. Painting on the World Wide Web: The Pumapaint project. In *IEEE International Conference on Intelligent Robots and Systems (IROS): Workshop on Web Robots*. Victoria, Canada, 1998.

[24] Organized by Ken Goldberg and Eric Paulos. In *IEEE International Conference on Robotics and Automation (ICRA): Workshop on Current Challenges in Internet Telerobotics*. Detroit, MI, 1999.

[25] Organized by Roland Siegwart. In *IEEE International Conference on Intelligent Robots and Systems (IROS): Workshop on Web Robots*. Victoria, Canada, 1998.

[26] Dirk Schulz, Wolfram Burgard, and Armin B. Cremers.   Predictive Simulation of Autonomous Robots for Tele-Operation Systems using the World WideWeb. In *IEEE International Conference on Intelligent Robots and Systems (IROS): Workshop on Web Robots*. Victoria, Canada, 1998.

[27] Gunter Niemeyer and Jean-Jacques Slotine. Toward force-reflecting teleoperation over the internet. In *IEEE International Conference on Robotics and Automation (ICRA)*. Leuven, Belgium, 1998.

[28] Kevin Brady and T. J. Tarn. Internet-based remote teleoperation. In *IEEE International Conference on Robotics and Automation (ICRA)*. Leuven, Belgium, 1998.

[29] K. Goldberg, J. Santarromana, G. Bekey, S. Gentner, R. Morris, C. Sutter, and J. Wiegley. A Telerobotic Garden on the World Wide Web. *SPIE Robotics and Machine Perception Newsletter* 5(1). March 1996. Reprinted in *SPIE OE Reports* 150, June 1996.

[30] Bobak R. Farzin, Ken Goldberg, and Adam Jacobs. A Minimalist Telerobotic Installation on the Internet. In *IEEE International Conference on Intelligent Robots and Systems (IROS): Workshop on Web Robots*. Victoria, Canada, 1998.

[31] Ken Goldberg, Steve Bui, Billy Chen, Bobak Farzin, Jacob Heitler, Derek Poon, Rory Solomon, and Gordon Smith. Collaborative Teleoperation on the Internet. In *IEEE International Conference on Robotics and Automation (ICRA)*. San Francisco, CA. April, 2000.

3 A Distributed Framework for Online Robots

Barney Dalton

3.1 Introduction

This chapter describes a framework that has evolved out of the University of Western Australia (UWA) telerobot project first started by Ken Taylor [14, 15, 16]. The framework has been designed to minimize the work required to connect a robot or other device to the Internet, while providing the basic functionality needed for a collaborative distributed system.

Most of the early World Wide Web (WWW) controlled robots [5, 14] used the Common Gateway Interface (CGI) [2] to interface between web browsers and the physical device being controlled. CGI processes are launched by a web browser in response to certain HTTP requests, and the CGI result is sent as the HTTP response. HTTP is a stateless request response protocol; it is simple to implement but it can have shortcomings. As it is stateless, state must be managed by the browser and the CGI process. This is usually achieved by passing extra identification information with each request, known as cookies [7]. The request response paradigm means that after a client request has been processed by a server, there is no way for the server to contact the client. The client must always initiate contact. This can be a problem when the client is interested in the state of a nonstationary remote process; to receive constant updates the client must poll the server at regular intervals. Polling is incfficient as requests must be made even when there are no changes and the server must handle these requests using resources to process them. Furthermore new information is only received with each poll, not as it becomes available.

The WWW has undergone major changes since the first online robots in 1994, and there are now new techniques available for robot control. The introduction of Java is probably the most significant as it allows code to be executed on the client side, whereas previously all code had to be in the CGI process. Java is being used by a number of online telerobotics projects, including the PUMA paint robot [15] and the NASA Pathfinder interface [1]. Using Java applets, a client can provide both a more sophisticated interface and use its own protocol to communicate with the server. The use of a different communication protocol means that the limitations of HTTP can be overcome. With a constant connection between client and server, both sides can communicate immediately when new information needs to be transmitted. For telerobotics, this ability for server initiated conversation is important as changes in the state of the robot and workspace needs to be uploaded to the client with minimal delay.

With each client running a Java applet and communicating with a server, the system can be called distributed. There are many issues associated with the design and implementation of a distributed system. The next section identifies the issues most relevant to distributed control of physical devices. Most of these issues are independent of the physical application, and they can therefore be addressed by a framework that can then be customized for specific applications. Section 2.1 presents a framework that has been implemented in Java to address these issues. The following sections discuss its application to the control of an industrial robot arm.

3.2   Design

An Internet controlled telerobotic system is likely to consist of a number of physical devices such as robot(s), cameras, sensors, and other actuators. Software to control these devices may require startup time that should be minimized by using permanently running servers. It is also likely that some devices may require different hardware and software configurations. The architecture of such a system is therefore likely to consist of a number of server processes that may be distributed across different machines. Added to this, the clients themselves will be in many different locations. The servers and clients are all part of the system and may all need to communicate with each other. In this context, both clients and servers are referred to as peers.

Writing servers can be complicated and time consuming. Not only does the software to perform the control and sensing need to be designed but a communication technique and protocol must also be chosen. This should enable communication among some or all of the peers in the system. The number, type, and location of peers may change over time. As communication may come from a number of peers, they need to be identified and authenticated. If the device being controlled cannot be shared freely, then its allocation to different peers needs to be managed. These and other requirements are similar for all servers and clients, and they can therefore be abstracted and provided as a framework.

Some of the issues and requirements that a distributed robotics framework must address include:

- Connectivity

- Authentication

- Addressing

- Resource allocation

- Time delay management

- Request delivery

Connectivity is the most basic requirement. For parts of a system to communicate with each other, they must be connected in some way. Potentially each peer may want to communicate with any number of other peers who are part of the system.

Authentication is essential to differentiate requests from different parts of the system. Both initial authentication and ongoing authentication may be required. Authentication typically consists of a challenge and response, where the actual response is compared to a expected response. Once authenticated a user's details can be retrieved. This may include information such as personal preferences, authority level, group membership, and personal information. Here the term user may apply equally well to a software server or agent or to a human user.

Addressing provides a mechanism by which one peer of the system can communicate directly with another peer by specifying the address of the recipient. The recipient my be a single peer, a list of peers, or all known peers.

Resource allocation is required for any resource that cannot be shared freely among all peers. This may be data files, device settings such as camera image size and quality, or physical devices themselves such as a robot.

Time-delay management is necessary for any control application. Currently the largest portion of the delay is caused by data traveling over a network. This is predetermined by the network path connecting parts of the system. Correct choice of protocols however, can help to minimize delay at the possible cost of reliability. Even though the delay cannot necessarily be controlled, it can at least be measured, providing estimates of information age, and therefore allowing predictions of the current state to be made.

For some devices, the order in which requests occur is important. These are usually commands that have some temporal or spatial relationship to the state of the device or its environment. For example, a relative command "move left" is issued on the assumption that the device is in a particular place. Any prior movement will render this command incorrect.

Reliable delivery and speed often have to be traded off against one another. To ensure reliable delivery, extra information must be passed backward and forward at the cost of speed. For important messages, reliable delivery must be chosen over speed, but there may be messages that do not require reliability and can therefore benefit from the increased speed gained from less handshaking.

3.2.1   Software Architecture and Design

The system architecture is based on Message-Oriented Middleware (MOM) [9, 11]. Peers connect to each other through a central router known as the MOM. This requires only n connections for n peers and also fits well with the Java applet security requirement that applets may only connect to the server from which they were served.

As shown in figure 3.1, the architecture is built up of a number of layers. The bottom layer is responsible for communication objects between connected processes and is known as the transceiver layer. Above the transceiver layer is the context layer. Contexts represent shared entities to which all peers may have access. Applications communicate by joining contexts and then sending/listening to messages through these contexts. Above the context layer are the client and presentation layers. All communication within the framework is encapsulated in objects known as parcels.

Parcels

Parcels consist of a list of keys and values. The key-value pairs are termed slots. The keys are strings, whereas their associated values may be any object type. Some example of slot types include address, sender, timestamp, priority, family, name, id, reference id, context, and content.

An address contains a list of peers to which the message is directed. If the address is omitted, then a broadcast is assumed. The sender slot identifies the creator of the parcel. Time stamps measure time-delay by indicating when a parcel was created. A time stamp may be significantly earlier than the time that the parcel is sent, as in the case where the system is busy and the parcel has to wait in queues.

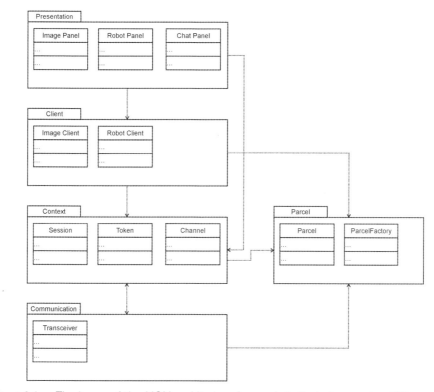

Figure 3.1:   The layers of the MOM architecture for a robot client shown using UML package notation. The presentation layer provides the user interface. The middle layers are explained in more detail later in this section. The bottom transceiver layer sends and receives objects over the network. Communication among the different layers uses parcels.

The priority of a parcel is an integer value expressing the urgency with which it should be delivered; higher values are given higher priority. If the priority slot is missing, the parcel is assigned a default priority. The family and name slots indicate the class to which the parcel belongs. The framework handles different types of parcels in different ways. The id and reference id are used for identifying parcels; a reply to a previous parcel would include the original parcel id as the reference id. The context slot identifies the context of the parcel (contexts are explained in detail in section 2.1). The content of the parcel is application specific and may contain any object, although in practice this must be of a type expected by the receiving peer.

```
Content-Type: multipart/related; boundary=xxxxBOUNDARYxxxx

xxxxBOUNDARYxxxx
Content-Type: text/xml
Content-ID: object

<?xml version=''1.0'' ?>
<parcel type=''parcel''>
  <properties type=''hashtable''>
  <key type=''string''>Family</key><value type=''string''>Request</value>
  <key type=''string''>Content</key>
    <value type=''hashtable''>
      <key type=''string''>Username</key><value type=''string''>barney</value>
      <key type=''string''>Password</key><value type=''string''>secret</value>
      <key type=''string''>Nickname</key><value type=''string''>b</value>
    </value>
  <key type=''string''>Name</key><value type=''string''>Login</value>
  <key type=''string''>Time</key><value type=''long''>946198913679</value>
  <key type=''string''>ID</key><value type=''int''>1</value>
  </properties>
</parcel>

xxxxBOUNDARYxxxx
```

Figure 3.2:   *A parcel converted to XML. This example is used for logging into the MOM at the start of a session. If the content was binary, then this would be included as a second part of the MIME parcel.*

An example parcel for logging into the system, converted to XML by the XML object transceiver (see next section for explanation), is shown in figure 3.2. The parcel is one used for logging into the system. The content is a name-value list of name, password, and nickname.

Transceiver Layer

The transceiver layer sends and receives objects through a connection with another transceiver. The other transceiver may be in the same process, a different process, or on a separate machine connected via the network. If a pair of transceivers are connected over a network, they must use a protocol to exchange objects.

Currently two implementations of network transceivers have been completed. One uses XML [17] as its communication protocol and the other delegates the task to Java's Remote Method Invocation (RMI) [10].

The XML-based transceiver is most useful for Java applets and cross-language use, and it was developed as a result of problems with RMI support in standard web browsers. Internet Explorer does not support RMI by default, and although Netscape Navigator supports RMI, there are problems with using RMI callbacks (see section 3.2 for discussion). XML is becoming a well-known standard for cross-platform exchange of data, and therefore it was chosen over a more concise custom language. To send an object as XML, the transceiver must know how to convert it into an XML representation. This is achieved by applications registering an XML writer that will perform this task. The XML writer may in turn recursively call other writers to serialize contents of the object. Each field of an object is written as an XML tag consisting of the name of the field with an attribute representing its type. XML is only designed for exchange of textual data and does not allow for the inclusion of binary data. To include binary data from an object, such as an array of bytes, the multipart/related MIME [8] type is used, binary parts are replaced by a cid tag in XML, and the binary part is included as a separate attachment. An example of a serialized parcel for logging into the system is shown in figure 3.2.

There are a number of disadvantages to the use of XML. First, new data types must have a defined XML translation, which needs to be registered with the MOM. Second, XML is extremely verbose and can therefore increase bandwidth requirements. This could be reduced by compressing the data before sending, but at the cost of extra processing time. XML is fast becoming an accepted standard, however, and there are many class libraries for parsing and generating it. It also has the advantage that it is human readable which makes it easier to discover errors.

The RMI implementation is useful for intra-Java communication within a local network. Where higher bandwidth is available, there are no firewalls, and there is control over client software. Because all the serializing and deserializing of objects is handled by RMI itself, no data format needs to be defined. The requirement is that RMI transceivers implement a remote transceiver RMI interface. As objects must be passed in both directions, once a client successfully obtains a reference to a remote transceiver, it must register itself with the remote transceiver so that it can be "called back."

Other implementations of transceivers are possible. An early version [4] used a structured binary protocol to send and receive data over sockets, but it was found to be too inflexible to changes in data structure. A CORBA implementation is planned. This would maximize cross platform and language interoperability. There are still problems with CORBA support in standard browsers however so there is little advantage to be gained at the moment. As discussed in section 3.2, not all clients are able to create socket connections on unprivileged ports. For these clients the only viable communication technique is to use HTTP tunneling. An HTTP tunneling transceiver would communicate with a Java servlet on the server side to relay parcels to the MOM.

Context Layer

The context layer provides the endpoint for parcels within the framework, therefore all parcels must be related to a particular context. Contexts are addressed in a similar way to URLs. They are organized in a tree structure relative to a single root context. Contexts can be managed by a manager specified at their creation. The manager is consulted to accept or reject all requests to the context. All contexts support the same basic actions, which are join and leave. To join a context a peer must have joined its parent first.

There are three main types of context: session, token, and channel. Sessions are like directories and are placeholders for a group of related contexts. Tokens are used for resource management and can be grabbed and released by peers. Access to tokens is determined by their manager which is specified when they are created. Channels are used by peers to exchange messages. Any message sent to a channel is forwarded to all peers in the address field or broadcast if the address field is missing. Some contexts in an example configuration are shown in figure 3.3.

To join a context, a peer first joins the context's parent session. A group of contexts can easily be managed by a single session parent, therefore which allows increasingly fine grained access control for each step down the context tree.

When the MOM is first started, only the root session exists — a configuration file may specify a manager that restricts creation of particular contexts to certain peers. Peers join the root session and then try to create or join appropriate contexts. For instance, a server that controls a robot might create a robot control token and a robot channel for exchanging commands and information about the robot.

Tokens play an important role as resource managers in the framework. Only one peer may own a token at any one time. Exactly how the token is managed, is determined by the creating peer. The robot token might be made immediately available, for example, if a peer has a higher user level than the current owner, or if the current owner has not renewed the token for a specified maximum time.

Contexts are implemented as master and proxy objects. The master object

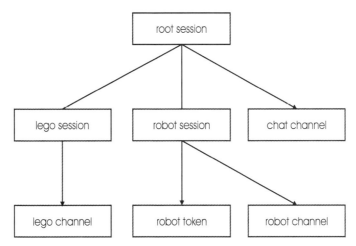

Figure 3.3:   *Contexts are organized in a tree structure similar to the structure of directories and files. Each type of context has a different purpose. Sessions are used as placeholders for logically related contexts, channels are used by peers to exchange messages, and tokens are used for resource control.*

resides in the MOM, while each peer uses proxy objects that point back to their respective master. All actions performed on proxy objects are relayed to the master object in the MOM. For instance, a "send message" action invoked by one peer on its chat channel context is first packaged as a parcel and sent via the peer's object transceiver to the MOM, where it is received by an object transceiver and forwarded to the root context. The destination context is examined and the parcel forwarded to the appropriate master context — in this case the chat channel. As the requested operation is a channel send, it is then queued and sent to each of the peers in the parcel's address list. The MOM resends the parcel using the object transceivers for each peer.

Client Layer

To use the framework in an application, references to the proxy objects are used for all operations. Initially, a session root object is created by specifying the address of the MOM server as a URL. Once a session root has been obtained, then other contexts can be found using a lookup or created if they do not exist and the manager allows it. Peers can then join contexts. Once joined, peers receive parcels for each context. Changes in context state are sent to peers as events which are a type of parcel. All contexts produce events when peers join and leave. Channels produce events when a message is received, while tokens produce events when they are grabbed or released. Sessions produce events when child contexts are created or deleted. The management and receipt of context events is usually handled by the client layer. In a simple application, the client layer may be omitted and events handled directly by the presentation layer.

3.2.2   The MOM in Practice

The MOM framework has been used to control a robot arm via Java applets distributed using a web server. The full system consists of the MOM, a robot server, an image server, and any number of Java applets dynamically downloaded by users. On system startup, the servers log into the MOM and create a number of contexts. The applet clients then join and subscribe to these contexts to control and observe the robot.

The robot server creates a robot session, under which there is a robot channel and robot token. The token is created with a manager who will only allow change of ownership if the requesting user has a higher access levels, or if the token ownership has not been renewed for three minutes. Robot commands are sent to the server over the robot channel and new robot states are broadcast by the server over the same channel. The server will only accept robot commands from the current owner of the robot token.

The image server creates a camera session under which it creates a channel for each camera. Commands to change image size and quality are received by the server over these channels. The server also uses these channels to broadcast information about new images. Image information includes the URL of the image, size, and calibration matrix (for those cameras that are calibrated). The image server subscribes to the robot token and only allows the robot token owner to change image specifications.

The client applets join all the robot and image contexts as well as an unmanaged chat channel. Using the robot contexts, control of the robot can be gained or released, robot commands can be sent, and command results and robot states can be received. Using the camera contexts new images can be requested and received, and image size and quality can be changed. Finally, the chat channel can be used to talk to other users as all are subscribed to the same unmanaged chat channel.

This shows just one example of the MOM framework in action. It could equally well be applied to a mobile robot or a simpler device such as a pan/tilt camera. This is not the only configuration for robot control. For example, it may be desirable to multitask the robot in different workspaces; in this case, a token could be created for each workspace and the robot server would then use those tokens to control robot access instead of the single token currently used.

It would also be easy to integrate agents into the system that could aid the user in planning tasks or even supervise robot motion. For example, a visual servoing agent could subscribe to the robot and image contexts. The user would specify two objects to align and would give control of the robot token to the visual servo agent. The visual servo agent would then keep moving the robot and analyzing new images until the desired state was reached, at which point it would return robot control to the user. The only change to the system would be to include the visual servoing option in the applet client and, of course, to create the visual servoing agent!

3.2.3   Interface Design

Most development work has been concerned with the core framework, and hence the current user interface is only a rudimentary implementation. The interface uses some, but not all of the facilities made available by the framework. As shown in figure 3.4, the interface has four main parts: a robot panel, a camera panel, user chat, and console area.

The robot panel lists any robot commands that the user has submitted and allows new ones to be added and edited. Commands can have a number of states and representations, but currently only commands based on a simple syntax are used. Commands are queued and can then be sent once the user is ready to execute them. The status of the commands can be one of the following: waiting to be sent, sending, executing, finished, or error. If an error occurs during the execution of a command, then the details are made available via the console. Currently there is no way of stopping a set of commands once they have been sent, but this could be implemented using a high-priority stop message delivered from user to robot server. As the robot moves, new poses are broadcast over the robot channel. The most recent pose is shown underneath the robot command list, and the path is drawn on the current camera image.

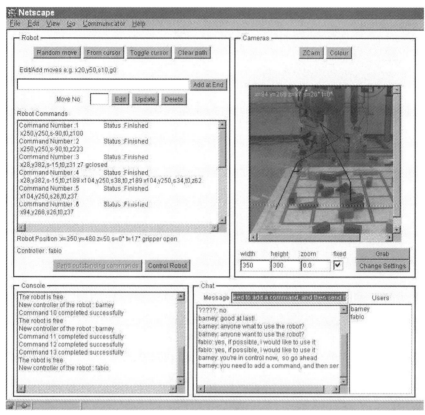

Figure 3.4:   The applet interface used to move the robot. The interface is divided into four parts.
The top left corner refers to the robot and the top right corner shows images taken by
the cameras. The console in the bottom left corner is used for returning information to
the user and the chat panel in the bottom right enables users to communicate with each
other.

The camera panels provide the user with the latest images of the workspace.
Different camera views can be selected using the buttons above the image. Each
camera has a corresponding channel context over which new image URLs are
broadcast. Commands to change image size and quality are carried over these same
channels. For calibrated cameras various objects can be overlaid. These include the
last n positions of the robot and an augmented reality cursor for specifying robot
pose. It is also possible to overlay wire-frame models of the blocks, although this
is not active in the current interface.

The console area provides a single visible place to show all error and status

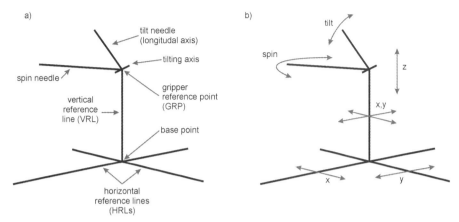

Figure 3.5: *The augmented reality cursor used as an alternative way to specify robot commands.*
Each element of the cursor can be used to specify a different degree of freedom.

messages. It is used as an alternative to the Java console available in most browsers, as most users either do not know how to show the console or, even if they do, rarely look at it. Robot errors and any unexpected communication errors are all displayed in the console. The chat area provides a basic chat interface. A single text field is provided for typing messages; on pressing "return," messages are broadcast to all users. Received messages are echoed in the text area below.

Robot commands follow a simple scripting syntax, allowing multiple way-points to be specified in a single command. These commands can be typed directly or generated from an augmented reality cursor displayed on the camera images (see figure 3.5). The cursor maps the five degrees of freedom of the robot to the two degrees of freedom of the image. Each line in the cursor represents a degree of freedom. By dragging part of the cursor, the user can manipulate each degree of freedom separately. To generate commands from the cursor, the user moves it to the required position and presses the "from cursor" button. The command is then pasted to the edit window where the user can make changes before queuing and sending it to the robot. The augmented reality cursor can also be used to measure position and orientation of objects in the image. For more details of the cursor and its application see [3].

3.2.4   Robot Hardware

The robot is an IRB 1400 made by Asea Brown Boveri (ABB). It is an industrial spot-welding robot with six degrees of freedom. The robot is controlled by an ABB S4 controller, running the ABB Baseware operating system with a RAP module for SLIP communication. The controller also has an additional digital I/O card that is used to control the gripper. The gripper has parallel jaws that are either open or closed.

Detection of contact and collisions is crude, as the only method available is the detection of a current overload in one of the joint motors. This cutoff value cannot be changed. To minimize collision forces, the robot is operated at 150mm/s, a fraction of its maximum speed.

The robot is restricted to operate in a 600 x 600 x 450 mm cuboid above the table. Orientation is also restricted to an inverted cone with an apex angle of 90 degrees. The orientation is further limited so that an imaginary line drawn through the gripper jaws is always in a plane parallel to the surface of the table. This intentionally reduces the degrees of freedom to five as the sixth is of little use for block manipulation. The two remaining orientations are termed "spin" and "tilt."

A serial link running the SLIP communication protocol connects the S4 controller to a networked computer. To execute commands on the controller, the ABB protocol RAP must be used. This is implemented in the ABB product RobComm.

3.2.5   Control

The S4 controller operates by running RAPID programs. A simple program is executed to move the robot. It first operates the gripper by switching the value of a digital output and then moves the robot to a target pose. To move the robot in a straight line, the following RAPID command is used:

```
MoveL   targetpos,vslow,fine,grippertool\WObj:=table;
```

This command means, move to pose "targetpos" with velocity "vslow", with tolerance of final position being "fine," using tool "grippertool" and workspace "table." The "targetpos" variable is currently the only variable that is changed for each move. For a more sophisticated application, control of the gripper speed would also be important. Currently, the speed is chosen as a balance between collision forces and path execution time.

At the suggestion of a user, an additional step was added to a basic move. If the gripper is moving down, the horizontal path is performed first; conversely, if the gripper is moving up, the vertical path is performed first. This reduces the chance of accidental collisions while executing the path since as the gripper is kept as far from the table for as long as possible.

During execution of a move, a large number of warnings/errors can occur. They may be from the request itself, from communication problems between the PC and robot controller, or there may be problems associated with the physical move. Errors in requests may be due to workspace boundaries or incorrect syntax. Typical physical problems include singularities, joint limits, and joint torque overloads. If the error occurs during the physical move, then the S4 controller provides an error code and some explanatory text. Often this text is quite technical and of little use to a users with little or no knowledge of robotics. For common errors, this explanation is replaced with a more understandable message.

3.3   The Setup

3.3.1   General Description

Image feedback is provided by a number of Pulnix TM-6CN cameras connected to a single Matrox Meteor II framegrabber. New images are taken whenever a new robot state is received on the robot channel, or when a user explicitly requests that a new image be taken. The position of the cameras has been varied over time, but the current positions are: one camera looking down the X axis; one looking down the Y axis; one looking down on the table; and one on the third joint of the robot arm. The X and Y axis cameras are calibrated. Calibration is performed by recording the position of the robot in camera images for twenty different robot poses, the camera matrix is then obtained by a least squares fit of the data. The calibration process is run as a Java application that joins the MOM system in much the same way as the user interface clients.

The MOM and image server run on a Pentium II with windows NT. The only extra card required is the framegrabber. For historical reasons, the robot server runs on a different NT machine, but it could probably be run from the same machine as

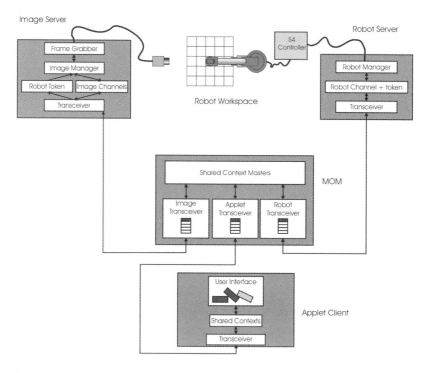

Figure 3.6:   The complete system.

the MOM. The SLIP link is run on a PC with Linux. Figure 3.6 shows the complete system.

3.3.2   Experience and Statistics

Previous experience and statistics with the CGI system have been reported in [13, 14, 15, 16]. The new MOM system has only been tested for short periods of time since as there are still a number of problems to be overcome. To check the network and Java capabilities of clients, however, a number of network tests using Java applets have been performed.

Network Tests

These network tests were designed to answer a number of questions. First, would users be able to connect to ports other than HTTP and, second, what sort of network performance could be expected?

Table 3.1: Connection's made from remote hosts. Just over 13% of clients failed to make socket connection back to the server on an unprivileged port. The reason for UnknownHostException is not understood, as the client would have to have been capable of resolving the server's name to download the test applet in the first place

Success			1162
Failures	java.net.UnknownHostException:	64	
	java.net.NoRouteToHostException: Operation timed out	56	
	java.net.ConnectException: Connection refused	39	
	java.net.NoRouteToHostException: Host unreachable	12	
	java.net.SocketException:	6	
			177
All			1339

The first question is whether a client can connect from an applet back to an unprivileged port on the server? For many networks, firewalls place restrictions on which ports can be used. If a client cannot connect on other ports, then the only communication technique available is HTTP tunneling. Table 3.1 shows the results for 1,339 unique hosts that downloaded the test applet; of these 87 percentage were able to make a connection. The remaining 13 percentage showed a range of errors, most of which probably can be attributed to firewalls. The cause of the "UnknownHostException" is not fully understood, as the client was able to resolve the host name to download the applet in the first place. It has been suggested that this is due to proxy configurations; where all DNS requests are handled by the proxy, straight DNS lookups within the Java VM therefore fail.

Figure 3.7 shows the time taken to establish an initial connection. The median time is 0.6 seconds with a third quartile of 1.4 seconds. This time is relatively insignificant if only performed once, but if reconnection is required for each request, as is the case with HTTP 1.0, then this can start to affect on the overall response time of the system.

To test the average network characteristics once a connection was established, the test applet sent and received fifty packets of a fixed size. The size of packet for each test was chosen at random and ranged from 1 to 10,000 bytes (figure 3.8 shows the results). Each point represents the average round-trip time for a particular connection; in all, almost 1,700 connections were recorded from 1,339

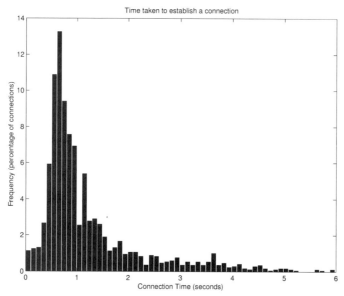

Figure 3.7:   *Time taken to establish a connection between a client applet and server at UWA. The distribution has a median of 0.6 seconds and a third quartile of 1.4 seconds. Connections should therefore be held open wherever possible to minimize waiting time.*

unique addresses. The graph shows that round-trip time appears to be independent of message size until the 1,000 byte range. This may be caused by all connections being carried in part by Ethernet which has a maximum transmission unit (MTU) of 1,500 bytes. This suggests that it may be better to send more data less often (as is implemented in TCP via Nagles algorithm) to achieve minimum average delay. It also shows that optimizing the XML protocol would have little effect on round-trip time provided that total message size is less than 1,000 Bytes. These tests give an approximate idea of average network characterizes of users. Further tests are being performed to obtain a more precise measure.

The above tests were carried out using TCP. Some tests using UDP were also tried, but there is a bug [6] in the Internet Explorer virtual machine that makes receiving UDP datagrams impossible in untrusted applets.

Results and Experiences

The system has been run and made available for some test periods. These have

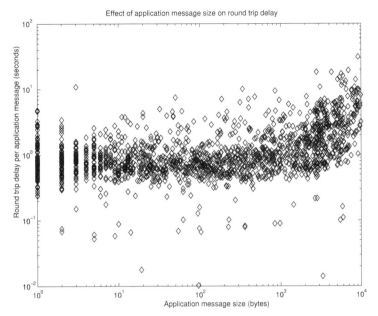

Figure 3.8:   *The effect of packet size on roundtrip delay. The overall trend is for round trip time to increase with packet size as would be expected. There appears to be minimal increase, however, until packet size reaches 1,000 bytes*

established that the applet works under Windows 95/98/NT, Linux, and Mac OS using Internet Explorer or Netscape Navigator versions 4 and higher. Users are able to move the robot and observe the status of commands, via the command status window, the overlaid robot path, and workspace images. The chat interface also provides a chance for users to collaborate, ask for help and exchange ideas. The collaboration capabilities of the framework have not been exploited by this first interface, but this is an area where it can provide significant improvements over the CGI version.

Some users report problems, although their platform and browser version correspond to those known to work. These problems are hard to debug as the problem machine is remote and unknown. This has led to the development of a remote logging technique that sends log information back to the sever using CGI instead of writing to the local error log. It is hoped that this will provide more information on client problems.

Although the idea of running Java applets in a client browser is attractive, there are many hurdles to be overcome. Developing a reliable applet is significantly harder than making a reliable Java application. The applet must obey all security restrictions, use the lowest common denominator of Java capabilities, and also be as small as possible to minimize download time. The size of the applet is reduced significantly using the IBM Java tool JAX; the final packaged JAR file was only 130KB — less than a minute download for most users. Another potential problem is that of Java memory use. The image server used 4MB of memory as C++ application, but once a Java front end was added using JNI, the usage increased to 16MB. Java applets can also use a significant amount of memory that for lower end client machines may render them unusable. Some Java implementations also have stability problems. The Virtual Machines (VMs) shipped with the Netscape and Internet Explorer browsers, although fine for lightweight, periodic use, prove less stable when operated for longer periods, often resulting in the browser or even the complete machine locking up. A possible solution to this problem is the use of the Java plug-in, although this requires a download of the Java runtime environment. The plugin is over 12MB which is a significant time commitment for users with slow modems. Web users are notoriously fickle in their attention span, so any extra requirements such as this are likely to reduce the use of the system.

Many Java applet problems come down to security issues. The problem of receiving UDP packets in Internet Explorer is a security manager issue that can be resolved by signing an applet and asking for higher network rights. Similarly, problems with the use of RMI callbacks in applets are caused by security exceptions, again signing the applet can solve the problem. Up until recently, however, no single standard has existed for signing applets, and therefore two separate, inconsistent systems have evolved for each of the major browsers. The Java 2 platform introduces a signing standard, but the default Java VMs of current browsers only support JDK 1.1. To use the new signing standard, therefore, the Java plugin must be used. Due to these signing complications, it is preferable to avoid API calls that require extra security so that applet can operate untrusted.

3.4   Conclusions and Future Work

The framework has been shown to provide the basic services required for a distributed robotic system operating over the Internet. The services include connectivity, user authentication, addressing, resource management, time delay estimation, and ordered request delivery. These services are provided through the abstraction of shared contexts such as sessions, channels, and tokens. The configuration and behavior of these contexts is determined only at run time and is specified by other parts of the system that connect to the central MOM.

The framework uses a hub-based architecture that consists of a central server (the MOM) and any number of peers connected to it. Thus architecture minimizes the number of connections in the system, while also adhering to the security restrictions of untrusted Java applets.

Application of the architecture has been successfully demonstrated for controlling a robot over the Internet. The robot, an industrial ABB manipulator, is positioned over a table with a number of cameras distributed a round the workspace. Users can use images from the workspace along with an augmented reality cursor to construct commands to send to the robot.

The framework provides a good basis for developing distributed telerobotic applications. There is nothing in the framework that is specific to robotics, and it may well be useful for other distributed applications. This also means that other distributed frameworks not originally intended for use in robotics may be able to be applied to telerobot applications. For example, Sun's Java Shared Data Toolkit provides many of the same features of channels, tokens, and sessions. JSDT requires applets to be signed, however, and because the JSDT libraries take up 700K, they also add significant size to the downloaded applet. Distributed applications is an area of intense research and there may be other projects that can provide useful contributions.

Although the system has been tested for short periods, it has yet to reach the stability of the previous CGI system. Therefore, immediate work will concentrate on further testing and bug fixing of the current system. Given that the design is a framework, it should be applicable to other applications. Some applications planned include the control of a small Lego Mindstorms mobile robot and a simple pan/tilt camera. It is envisaged that if all these robots were controlled from the same MOM, they could be used together to form a collaborative system with multiple robots and multiple users.

Development of the robot-arm system could proceed in many directions, although increasing the level of robot commands is a priority. This has been investigated in an interface that allows a user to interactively model objects in the environment. Commands could then be generated as objects instead of position.

This could be implemented by introducing a model server and a model channel, where updates and queries to the model would be performed using the model channel. Written access to the model could be restricted to the owner of the robot token or new model tokens could be introduced.

References

[1]   Paul G. Backes, Kam S. Tao, and Gregory K. Tharp. Mars pathfinder mission Internet-based operations using WITS. In *Proceedings of IEEE International Conference on Robotics and Automation*, p. 284 – ed.291, May 1998.

[2]   The common gateway interface. http://www.w3.org/CGI/.

[3]   B. Dalton, H. Friz, and K. Taylor. Augmented reality based object modelling in internet robotics. In Matthew R. Stein, ed., *SPIE Telemanipulator and Telepresence Technologies V*, vol. 3524, p. 210 – 217, Boston, 1998.

[4]   B. Dalton and K. Taylor. A framework for internet robotics. In Roland Siegwart, ed., *IROS Workshop Robots on the Web*, p. 15 – 22, Victoria B.C. Canada, 1998.

[5]   K. Goldberg, K. Mascha, M. Genter, N. Rothenberg, C. Sutter, and J. Wiegley. Desktop teleoperation via the world wide web. In *Proceedings of IEEE International Conference on Robotics and Automation,* May 1995.

[6]   Derek Jamison. Bug: Datagram socket causes security exception. http://support.microsoft.com/support/kb/articles/Q186/0/32.ASP, 1998. Microsoft Knowledge Base no Q186032.

[7]   D. Kristol and L. Montulli. HTTP State Management Mechanism. RFC 2109. Technical report, Internet Engineering Task Force (IETF), February 1997. Available at http://www.ietf.org/rfc/rfc2109.txt.

[8]   E. Levinson. The MIME Multipart/Related Content-type. RFC 2387. Technical report, Internet Engineering Task Force (IETF), 1998. Avaiable at http://www.ietf.org/rfc/rfc2387.txt.

[9]   Robert Orfali, Dan Harkey, and Jeri Edwards. The essential client/server survival guide, chapter RPC, Messaging, and Peer-to-Peer, Wiley computer publishing, 1996.

[10] Java remote method invocation. http://java.sun.com/products/jdk/rmi/index.html.

[11] Michael Shoffner. Write your own MOM. http://www.javaworld.com/javaworld/jw-05-1998/jw-05-step.html, May 1998. Internet Publication - Java World.

[12] Matthew R. Stein. Painting on the world wide web: The PumaPaint project. In Matthew R. Stein, editor, *SPIE Telemanipulator and Telepresence Technologies V*, vol. 3524, p. 201 – 209, Boston, November 1998.

[13] K. Taylor and B. Dalton. Issues in Internet telerobotics. In *International*

Conference on Field and Service Robotics (FSR 97), p. 151 – 157, Canberra, Australia, December 1997.

[14] K. Taylor and J.Trevelyan. Australia's telerobot on the web. In *26th International Symposium on Industrial Robotics,* p. 39 – 44, Singapore, October 1995.

[15] Ken Taylor, Barney Dalton, and James Trevelyan. Web-based telerobotics. *Robotica* 17:49 – 57, 1999.

[16] Kenneth Taylor. Web Telerobotics: Reducing Complexity in Robotics. *Ph.D. thesis*, University of Western Australia, 1999.

[17] Extensible markup language (xml) 1.0. http://www.w3.org/TR/1998/REC-xml-19980210.

4 Online Robots

Hirohisa Hirukawa, Isao Hara, and Toshio Hori

4.1 Introduction

When the World Wide Web is seen as an infrastructure to build a robotic system on, its attraction is threefold. First, web browsers can be a good human interface in a robotic system because they can display various media, including hypertexts, still images, movies, sounds, and 3-D graphics as well as handle interactive operations on the media. An example would be a manipulation of a 3-D object described in VRML by a mouse. Second, Hypertext Transfer Protocol (HTP) can be a standard communication protocol of a telerobotic system since robots connected to the Internet can be accessed from any Internet site via the protocol. Third, it becomes possible to use various robotic hardware/software distributed on the Internet together to accomplish a single mission. Therefore it is natural to consider the development of robotic systems on the Web; here this concept is called *webtop robotics.*

To exploit the first point, the authors have been investigating VRML browsers and Java3D API as a 3-D graphics display and API for geometric processing that are the major building blocks of interactive motion planners based on an off-line simulation

The specification of VRML97 includes the function to process interactive input operations on objects. This function is specified as a subset of sensor nodes in VRML97, which includes CylinderSensor, PlaneSensor, and SphereSensor. The CylinderSensor maps pointing device motion into a rotation along the axis of a virtual cylinder, the PlaneSensor into a translation on a plane, and the SphereSensor into spherical rotation around the center of a virtual sphere. The pointing device can be a mouse for 2-D input and a wand for 3-D, so the specification is rich enough for 2-D or 3-D input. But teleoperation systems need 6-D input devices in general, like a master manipulator, with 3-D for a translation of an object and 3-D for a rotation. VRML97 browsers must be enhanced in some way to handle 6-D input.

Considering an interactive motion planner for mechanical assembly, the planner needs to handle the kinematics of objects in contact. VRML97 can handle lower pairs by the sensor nodes described above, since pointing device motion is mapped into 1-D or 2-D motion of lower pairs. That is, a revolute pair can be processed by the CylinderSensor, a planar pair and a prismatic pair by the PlaneSensor, a spherical pair by the SphereSensor, and a cylindrical pair and a screw pair by the combination of the CylinderSensor and PlaneSensor. The specifications of VRML97 do not include how to manage higher pairs, since it is not easy to process higher pairs efficiently. A VRML97 browser has been enhanced to handle higher pairs, whose details will be described later.

An interactive motion planner on Java3D API has also been developed. With a small plugin, Java3D can be used on web browsers as well. Why does one need the implementation on Java3D other than that on VRML browsers? Because VRML browsers are very convenient for small-scale enhancement, but not for large-scale due to the lack of API for geometric processing. Several implementation examples on Java3D will be also shown.

Now consider the second and third points together. When robots are connected to the Internet, each pair of an operator site and a robot is still considered to be a telerobot system, but it is now slightly different from the conventional concept. The system now shares communication lines with other resources, and every system has the potential to communicate with the other systems on the network. These new systems connected to computer networks can be called *networked robot systems*.

There are, however, several difficulties in forming such a robot network on computer networks. For example, we must know what kind of robots are connected to the network and where they are, and note that the factor of the interoperability of robots is neglected and controlling interfaces among them are incompatible. Building robot systems on the web is a promising solution to these problems.

A further difficulty of networked robot systems is the unpredictable time delay of the Internet. To overcome this, one option is to leave lower level control to the robot-side controller and concentrate on higher level control. Thus one considers a remote robot as an object, because the robot controller in the system can hide the precise structure of the robot from the human operator, and one needs not know about the robot's internals in detail.

This is similar to the object-oriented approach in computer programming where everything in the system is mapped onto an object and communication between objects is expressed as message passing. The object-oriented approach has three characteristics: data abstraction, inheritance, and dynamic binding. With data abstraction, the interfaces for managing an object's status are abstracted, so the internal data and implementation of information processing methods in the object, the structure of the object's class, are encapsulated and hidden from users of the object. If the class structures of existing objects have been carefully designed, a newly created object may only need to inherit them and add its own special data structure and methods. The object/method called during operation may be not statically bound at the compile stage but dynamically bound according to the situation.

These ideas can be introduced into the implementation of networked robot systems. Users need not know the robot's internals to control it but only its control interface. If several standardized classes and interfaces for controlling robots are designed and implemented, a robot newly connected to a computer network needs only inherit them and add its own specific functions internally. So the modularity of robot control programs will be increased and their reusability will also increase as a result. Users can select a robot at run-time according to what they want to do with the robot. Therefore, looking at the communication layer of a networked robot system, one can map the architecture of the object-oriented approach onto that of a networked robot system, and one can say that a robot is an object and networked robots are distributed objects.

The concept of distributed objects is also interesting for the development of a robotic software library distributed on the Internet. Robotic systems need various elemental technologies, including vision, force sensing, motion planning, and control; and the required specifications of each technology can be milder when the systems are equipped with more elements. But it is hard to develop several elemental technologies by a single stroke. Therefore, there is a great demand for a robotic software library, which enables one to share various up-to-date technologies. Very few software products, however, have been made public in the robotics community so far.

From these observations, the authors propose to implement networked robot systems on the Web that is enhanced by CORBA (Common Object Request Broker Architecture). CORBA [11] is a specification for distributed object computing so there are many implementations on diverse operating system platforms. Any CORBA-compliant implementation can communicate with the others as long as they follow the specification. To establish interoperability among different implementations, the Internet Inter-ORB Protocol (IIOP) is specified to bridge communication. The Interoperable Object Reference (IOR) is a standardized object

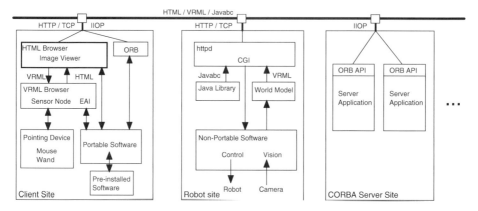

Figure 4.1:   Configuration of networked robot system on the Web.

reference format and can be used between different CORBA compliants to identify a specific object. IOR can be stringified and stored in a plain file so one can open it, for example, on the web. Moreover, CORBA defines the interface repository, which provides persistent storage of interface definitions of objects to support dynamic invocation and the naming service which provides a mapping from object names to object references.

The latency of communication over a computer network is unpredictable, so one can not expect real time processing in the networked robot systems. If CORBA degrades the system's performance much further, however, it is inappropriate to use it. So its performance has been evaluated in terms of the data transfer latency of a CORBA implementation. Implementation examples of distributed robotic systems will also be presented.

The next section revisits the anatomy of the WWW system together with CORBA and shows how to map robotic systems on it.

4.2   Design

4.2.1   Software Architecture and Design

A mapping of robotic systems on the WWW system is shown in figure 4.1. Here, the operator site is supposed to be a human interface to control a robot in an online fashion or to plan robot motions in an off-line way. The remote site, not necessarily in a remote place, is a server site on or near the robot.

At the operator site, the HTML browser can display 2-D images, including moving pictures by JPEG/MPEG format and the VRML browser 3-D graphics with interactive operations on them by a mouse/wand. These functions are able to be extended by a network of portable software like Java/JavaScript. Notice that these media and software can be transferred from the remote robot site on demand, and that no specific software is required to be installed at the operator site a priori. At the present implementation, the speed of Java codes is not comparable with that of the corresponding C/C++ ones even if a JIT compiler is used. When the speed of Java codes is not satisfactory, the critical part of the enhancement must be implemented by pre-installed software written in C/C++ at the operator site.

At the remote site, the httpd works to transfer images of the working environment of the robot, 3-D geometric models of the environment and the robot itself with its kinematics, and Java byte codes to enhance the functions of the operator site. The software to control the robot hardware and to sense its working environment by vision or other sensors need not be network portable because it is not sent to the operator site.

Sometimes it is necessary to make a direct TCP connection between the operator and robot sites. This does not break the security rules of popular browsers, like Netscape Navigator, because the rules permit a socket connection between a client site and a process on the server from which the current HTML or VRML came. This is not preferable, however, because general firewalls of Internet sites do not permit the socket connection from outside and such a connection suffers from unpredictable delay time of the Internet. The employment of CORBA is a better alternative, though IIOP has not yet been made transparent through the firewalls of most sites.

Within this general architecture, the following two types of teleoperation systems have been implemented [4,5]. From the viewpoint of motion command transfer, teleoperation systems can be divided into two groups. In the first group, mostly conventional ones, an operator directly controls a real robot and manipulates objects at a remote site. This group is called a *direct type*. In the second group, an operator plans robot task interactively on a computer simulation of a robot and objects, that is, in a virtual world, and the planed motions will be sent to control a remote real robot afterward.This group is called an *indirect type*. The indirect type has been developed mainly to cope with time delay on the

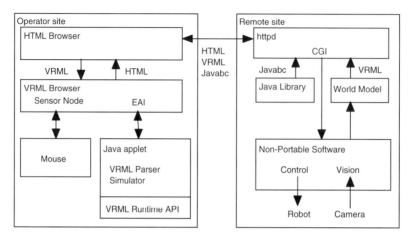

Figure 4.2:   Configuration of the indirect teleoperation system.

communication path, that is, the signal transmission path.

A teleoperation system has been developed for mechanical assembly of the indirect type and for a monitoring camera of the direct type. The configuration of a teleoperation system for mechanical assembly of the indirect type is shown in figure 4.2. The indirect teleoperations for mechanical assembly take place in the following order. First, the geometric model of the working environment of the robot is supposed to be obtained by a vision system at the robot site, and the model is converted to VRML97 format. A converter from a conventional geometric model to VRML97 format has been developed, and many converters from standard formats to VRML97 are also available. Next, the VRML file is sent to the client site by the operator's click on the web browser with Java byte codes to enhance the VRML browser to be an indirect teleoperation system. The main enhancement consists of a GUI converting 2-D motions of a mouse to a 3-D translation or 3-D rotation to manipulate an object, and collision detection between the manipulated object and the fixed objects. The collision detection is implemented by a Java porting of Rapid [2], which is originally implemented in C++.

It turned out that the speed of the collision detection by the Java code is about one-tenth of the original one, and the speed is not satisfactory when the number of polygons in the scene becomes thousands. So a system has been developed using the native code, which must be preinstalled at the client site a priori.

A serious problem of implementation is that the topologies of geometric

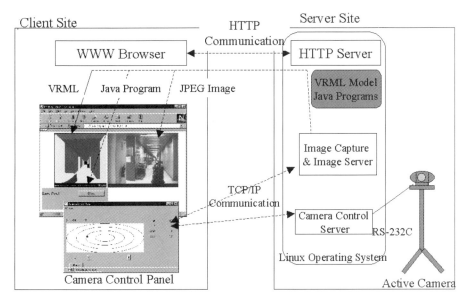

Figure 4.3:   Remote control of a camera.

models in VRML browsers can not be seen from a Java applet according to a specification of VRML97. As a bypass for this problem, a VRML file is read into the Java applet and parsed at the same time as the file is read by a VRML browser, and the mirror models are maintained in the applet and the browser. A subset of VRML runtime API on the applet has been developed for this purpose.

Another teleoperation system has also been developed for mechanical assembly of the indirect type, where the input mouse is replaced by a six DOF manipulator with force feedback capability [4]. In this system, interactive operations by the manipulator are handled by a Java applet instead of sensor nodes of VRML97 browsers, whose specifications do not include such operations.

The configuration of a teleoperation system for a monitoring camera of the direct type is shown in figure 4.3 with a snapshot of the web browser. Moving images from the camera are sent from the image capture server to the web browser in JPEG format on HTTP. The camera server also has a VRML model of its environment, and the snapshot from the current camera position is shown in the VRML browser.

Pan/tilt motions of the camera can be controlled by the GUI on a Java applet, and the control commands are transferred on a socket connection of TCP/IP. The

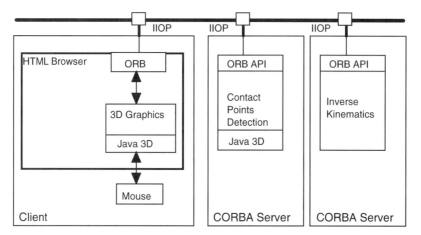

Figure 4.4:   Example of a distributed robotic system.

commands can also be transferred on HTTP to go through fire walls, but then it becomes difficult to control the velocity of the camera motions since the connection must be established every time the command is sent and the establishment is unstable due to the traffic on the Internet.

When the motion commands are input, the camera moves after some time delay. Then an operator tends to input more commands which causes the camera to overshoot the desired position. To avoid such operations, the graphics on the VRML browser moves first according to the motion commands with no time delay to clarify how much the operator has moved the camera position.

A distributed robotic system has also been implemented on the Web enhanced by CORBA, which is a distributed version of the indirect teleoperation system [6]. The configuration of the system is shown in figure 4.4.

A user moves the end-effector of a manipulator interactively by a mouse, whose 3-D graphics are built on Java 3-D API. Then, the motion command is transferred to the inverse kinematics server, and the corresponding joint angles of the manipulator are returned from the server. Next, the current positions and orientations of the links of the manipulator are sent to another CORBA server to detect contact points between the links and the environment, and contact points are returned if any. These CORBA servers can be located at any Internet in principle. This example of the distributed robotic systems illustrates how clients can use it on the webtop.

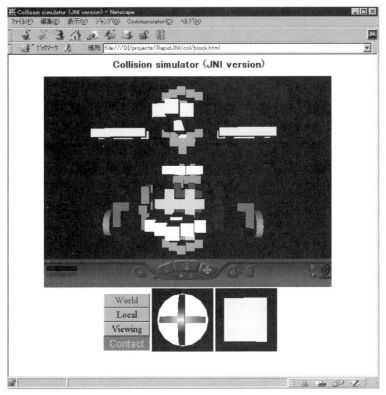

Figure 4.5:   Assembling blocks.

4.2.2   Interface Design

A snapshot of the indirect teleoperation system is shown in figure 4.5. Cosmo Player from Silicon Graphics Inc. is used as the VRML97 browser. The cube in the lower part of the browser is the GUI to input 3D translations, and the rings are to input 3-D rotations, both of which are also implemented on Cosmo Player. VRML97 browsers can handle lower pairs by sensor nodes, and a pointing device motion is mapped into a 1-D or 2-D motion of lower pairs.

A snapshot of the interface of the direct system was shown in figure 4.3. A snapshot of the distributed system is illustrated in figure 4.6. Users can move the end-effector of the manipulator interactively, and the system solves its inverse kinematics and checks the collision with the working environment. The line

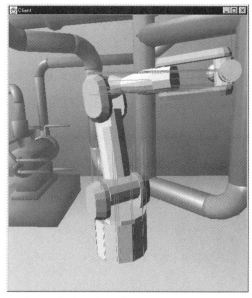

Figure 4.6:   Snapshot of a manipulator in a plant.

segments in the image show the normal vectors of the surfaces at the contact points.

4.2.3   Robot Hardware

The planned motions in the indirect teleoperation system can be executed afterward by a real manipulator like MHI PA-10 (see figure 4.7). An execution system has been developed under a separate project, but it is beyond the scope of this chapter. (see [10] for details). In the case of the direct teleoperation system, the robot is a movable camera and can also be a mobile robot like a Nomad (see figure 4.8).

4.2.4   Control

In the indirect system, the conventional controller can be used to control the manipulator without any extension because motions are planned in an off-line way and they are supposed to be executed afterward. Rather than this approach a way to enhance VRML97 browsers is needed for handling the kinematics of the higher pairs in the off-line motion planner.

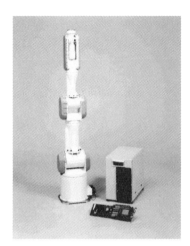

Figure 4.7:   MHI PA-10 manipulator.

Figure 4.8:   Mobile robot Jijo2.

To this end, the authors have developed the algorithms that can find possible motions of a moving object in contact with fixed objects, even if the moving object and the fixed objects compose higher pairs [3]. These algorithms are reviewed here briefly. Assume that the shapes of objects are polyhedra, that there is no friction among them, and that objects moves slowly. Then the constraint for the motions of an object imposed by contact with fixed ones can be reduced to the conjunction of the constraints at a finite number of contact points and can be given by

$$H_i^T \cdot V \geq 0 \text{ for all } i = 1, ..., n \tag{4.1}$$

where V: 6x1 is the screw or the translational and angular velocity of the moving object, $H_i = F_i^T \cdot J_i$, F_i: 3x1 the normal vector of the constraining face at the i-th contact point, J_i: 3x6 the Jacobian matrix that transform V to the translational velocity at the i-th contact point, and n the number of the contact points. The solution of equation 4.1 stands for the set of the possible velocity of the object at the current contact state, and it is a polyhedral convex cone in the screw space of the object.

When a motion command is input, our final mission is to find the feasible element of the input velocity, optionally with the repulsive force. This problem can be reduced to find the orthogonal projection of a point onto a polyhedral convex cone, which can be an application of some nice data structure that maintains the convex hull of a dynamic points set and can answer the query in $O(\log n)$ time (see [9]). But in this case, n is relatively small and the polyhedral convex cone may change drastically according to the input commands that invoke the expensive update and preprocessing of the data structure. So a naive but simple algorithm has been developed that can answer the query in $O(n)$ time or a few msec. We apply these algorithms to develop the interactive motion planner.

4.3   The Setup

4.3.1   General Description

A real manipulator has not been made public to anonymous users on the Internet; the off-line motion planner has been set up on the Web. The movable camera has been made public for sometime. Anonymous users could then take a look at the inside of our office while controlling the camera. The mobile robot was connected to the network several times, and it was controlled from remote sites. Though robots have not existed for a long period, the networked robot systems can be operated in principle on the Web anytime.

Here, the possible applications of such robots on the Web are discussed. One

Figure 4.9:   Snapshot of a virtual home.

possible application is tele-maintenance of the home. Examples of the telemaintenance tasks include checking to see if a gas tap is closed properly, a gas burner turned off, or the entrance door closed. Though such checking can also be done by a movable camera, we need a teleoperation system to correct the wrong situation if it exists.

Assume that a good mobile robot with a perfect vision system and a dexterous manipulator exists at home. (A snapshot of a virtual home is shown in figure 4.9). The rotating doors of the closet on the wall of the living room can be opened or closed by a mouse pointer, since a CylinderSensor is embedded along the axis of each door. A sliding door can also be moved, thanks to the embedded PlaneSensor.

Another possible mission is manipulating an electronic apparatus. Imagine the following scenario. One day, a guy realizes that he has forgotten to record a video of his favorite program when he is away from home. He rushes to a public Internet site and browses for his server at home. He can see the inside of his home through the vision of the robot at his home, and he navigates it to the front of his video

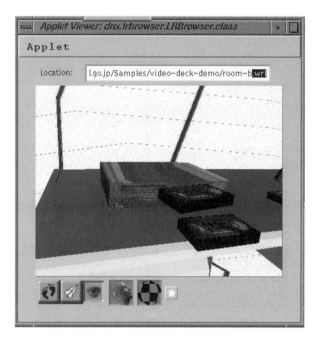

Figure 4.10: Inserting a cassette into a recorder.

recorder. Next, the geometric models of the recorder and a blank tape with their current locations are sent to his local site, and he tries to set the tape and record the video on the simulator. A snapshot of the inserting motion is shown in figure 4.10.

Finally, the obtained motion commands are sent to his home to complete his mission. While he is trying to insert the tape into the video recorder, the tape may be in contact with the recorder at some points, line segments, or facets. So the tape and the recorder compose a higher pair in general, which is handled by our manipulation simulator.

Another possible application of the system is 3-D virtual shopping, where a customer can look at merchandise visually from various directions while manipulating it. A snapshot of such virtual shop is shown in figure 4.11. It would be more interesting if images of a real shop were combined with its virtual model, where the images could be obtained by video cameras mounted in the shop. Future work includes how to fuse such real and virtual worlds to realize an enjoyable shopping system.

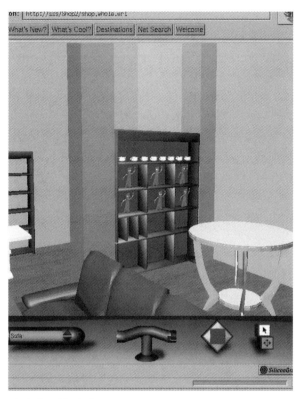

Figure 4.11: Snapshot of a virtual shop.

4.3.2   Experience and Statistics

Robots have been set up on the Web in the way described in the previous section, but unfortunately no significant statistics have been generated. Instead, an implementation of CORBA has been evaluated to estimate the performance of the networked robot systems.

The performance of an implementation of CORBA has been evaluated through the following experiment. The client sends a remote method call with 64 (or 128) double precision words (8 bytes per word) to the server, and the server returns the identical data to the client. This cycle is repeated 1,000 times, which is called one set. Ten sets are done for each combination of the client and server, and the best performance is chosen to be the representative result from ten sets and its average in a set is seen as the round-trip time of the call. The client is located at PC1, and the server is at the identical PC, another PC2 in a LAN, or at yet another PC3 in a WAN. The hardware layer of the LAN is 100BASE-TX and that of the WAN 6Mbps link. The number of hops of the LAN is 3, and that of the WAN 17. The specifications of PCs are shown in table 4.1.

ORBacus 3.0b1 is used as an implementation of CORBA. The experiment done to the combinations of the client and the server is shown in table 4.2 with results.

The unit of time is msec. From the second row of table 4.2, the overhead is 0.60 msec for C++/C++/64 words. This value is not small enough for applications like servo control of a motor, but acceptable for applications whose communication cycle is around 10 msec.

Table 4.1: Specifications of PCs

PC1	Pentium II 300MHz,160MB, NT4.0
PC2	Pentium Pro 200MHz x 2, 128MB, NT4.0
PC3	Pentium Pro 200MHz x 4, 128MB, NT4.0

There is no significant difference between the results of the combinations of C++/C++ and Java/C++ when LAN or WAN is used. The round-trip time is almost the same as that of ping command, and therefore one can know the rough value of overhead of CORBA by executing the command to the desired server. The round-trip time of the WAN is also acceptable for applications whose communication cycle is around 10 msec. So, one can say that a prerequisite of robot systems on the Webis as follows: The communication cycle of the computation is not shorter than 10 msec. This threshold may vary according to the round trip time of a WAN used for the distributed system.

Table 4.2:   Round trip time of CORBA

Path	Client	Server	64Words	128Words
Local	C++	C++	0.60	0.65
	Java	C++	1.48	1.90
LAN	C++	C++	3.62	4.40
	Java	C++	3.67	5.86
WAN	C++	C++	29.18	35.58
	Java	C++	28.18	31.05

4.4   Conclusions and Future Works

The authors have proposed the concept of webtop robotics. The results can be summarized as follows:

- The configuration of the WWW system was analyzed, and a mapping of robotic system onto the Web was presented.

- Teleoperation systems of both the indirect and the direct type have been developed as implementation examples of the concept as well as a distributed version of the indirect type.

- It has been claimed that networked robot systems can be considered as a distributed object system, and that robots on the Web can be enhanced by CORBA naturally.

Future work includes the development of the real applications described above with real manipulators and mobile robots.

Acknowledgments

The authors thank Susumu Kawamura and Ichitaro Kohara for helping with the implementation of the system.

References

[1]   K. Goldberg, et al., Desktop teleoperation via the World Wide Web. In *IEEE International Conference of Robotics and Automation,* pp.654–659, 1995.

[2] S. Gottschalk, M. Lin, and D. Manocha. OBB-Tree: A Hierarchical Structure for Rapid Interference Detection. *Proc. of ACM Siggraph*, 1996.

[3] H. Hirukawa, T. Matsui, and K. Takase. Automatic determination of possible velocity and applicable force of frictionless objects in contact. *IEEE Trans. Robotics and Automation*, vol.10, no. 3, pp.309–322, 1994.

[4] H. Hirukawa, T. Matsui, H. Onda, K. Takase, Y. Ishiwata, and K. Konaka. Prototypes of Teleoperation Systems via a Standard Communication Protocol with a Standard Human Interface. In *IEEE International Conference on Robotics and Automation*, pp.1028–1033, 1997.

[5] H. Hirukawa, T. Matsui, S. Hirai, K. Konaka, and S. Kawamura. A Prototype of Standard Teleoperation Systems on an Enhanced VRML. In *IEEE/RSJ International Conference on Intelligent Robots and Systems,* pp.1801–1806, 1997.

[6] H. Hirukawa and I. Hara. The Web Top Robotics, Workshop on Robots on The Web. In *IEEE/RSJ International Conference on Intelligent Robots and Systems*, pp.49–54, 1998.

[7] T. Hori, H. Hirukawa, T. Suehiro, and S. Hirai. Networked Robots as Distributed Objects. In *IEEE/ASME International Conference on Advanced Intelligent Mechatronics,* pp.61–66, 1999.

[8] M. Mitsuishi, T. Hori, H. Kuzuoka, Y. Hatamura, and T. Nagao. Computer Networked Cooperative Manufacturing. In *IEEE Int. Workshop on Robot and Human Communication,* pp.127–134, 1996.

[9] K. Mulmuley. *Computational Geometry–An Introduction through Randomized Algorithms*, Prentice-Hall, 1994.

[10] H. Onda, H. Hirukawa, and K. Takase. Assembly motion teaching system using position/force simulator. In *IEEE/RSJ International Conference on Intelligent Robots and Systems*, pp.9–16, 1995.

[11] http://www.omg.org/

Part II
Remote Mobility

5   Xavier: An Autonomous Mobile Robot on the Web

Reid Simmons, Richard Goodwin, Sven Koenig, Joseph O'Sullivan, and Greg Armstrong

5.1   Introduction

In December 1995, the authors began what they assumed would be a short (two to three month) experiment to demonstrate the reliability of a new algorithm that they had developed for autonomous indoor navigation [12]. To get a significant variety of commands to the robot over an extended period of time, web pages were set up that enabled users throughout the world to view the robot's progress and command its behavior. What they failed to appreciate, at the time, was the degree of interest an autonomous mobile robot on the Web would have. Now, 30,000 requests and 210 kilometers later, Xavier (figure 5.1) continues to fascinate people who have read about it in the popular press, stumbled across its web page, or found it through one of the many links to its web site (http://www.cs.cmu.edu/~Xavier).

Although the main research focus has not been the web-based aspects of Xavier, the experiment has taught several things about the interactions between remote users and autonomous robots. The main lessons are about robot reliability and the types of remote interactions that are useful, together with sociological anecdotes about people, technology, and the Web.

Xavier differs from most other online robots in that it is both mobile and autonomous (the Rhino and Minerva tour guide robots [1, 13] are similarly successful autonomous mobile robots with web-based interfaces). Mobility affects online robots because, at least until very recently, the bandwidth achievable by affordable wireless modems has been rather limited. Thus, real-time visual feedback and control is often difficult to achieve, especially if the workspace of the robot is a large area (such as a whole building) so that wireless coverage becomes a factor. Also, battery power is limited, so the robot can operate only a few hours per day.

Autonomy can help in reducing the bandwidth requirements for control but introduces problems of its own, particularly in the area of interactivity. People appear to prefer "hands-on" control, and they do not appear to experience the same type of immediate feedback with an autonomous mobile robot as they get with a teleoperated one. This is exacerbated by the limited up-time of the robot, which

Figure 5.1:   Xavier robot.

reduces the chances for people to see the robot operating when they happen to come to its web site.

Despite these challenges, the authors believe that Xavier has been a successful (if somewhat inadvertent) experiment in online robotics. In particular, it has given thousands of people their first introduction to the world of mobile robots. Judging by the feedback received, the overall response to Xavier has been extremely positive and people are generally impressed that robots can have such capabilities (in fact, many people have expressed the desire to have such a robot of their own).

5.2   Design

Robotic systems are typically intricate combinations of software and hardware. Usually, the hardware and software are tuned to the particulars of the environment in which the robot must operate. The Xavier system was developed to operate in peopled office environments. In particular, it must have reliable long-distance navigation capabilities. It also must interact with users who may be remotely distributed (e.g., office workers trying to ask the robot to do something). The following sections describe the software and hardware architectures of Xavier that support these capabilities.

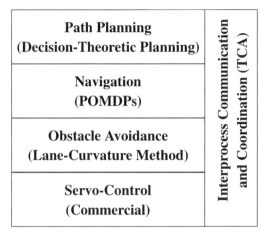

Figure 5.2:   The navigation system.

5.2.1   Software Architecture and Design

The Xavier navigation system is a hierarchically layered architecture (figure 5.2), consisting of servocontrol, obstacle avoidance, navigation, and path planning. Each layer receives "guidance" from the layer above and provides commands to the layer below. Each layer also filters and abstracts information for the higher layers, thus enabling them to operate more globally without getting swamped by data. While the navigation system and each of the individual layers have been described elsewhere [11], a brief overview is given here of the salient features of the system.

The *servocontrol* layer, which controls both the base and pan-tilt head, provides simple velocity and/or position control. It also provides feedback on command execution and position information, based on encoder readings. The servocontrol layer is primarily commercial software that comes with the hardware (in particular, the base and pan-tilt head).

The *obstacle avoidance* layer keeps the robot moving in a given direction, while avoiding static and dynamic obstacles (such as tables, trash cans, and people). It uses the lane-curvature method [8], which tries to find highly traversable lanes in the desired direction of travel, and uses the curvature-velocity method [10] to switch between lanes, avoid dynamic obstacles, and slow down in crowded situations. Both methods take vehicle dynamics into account to provide safe, high-speed motion (Xavier averages about 45 cm/sec in peopled environments).

The *navigation* layer is responsible for getting the robot from one location to another. It uses a partially observable Markov decision process (POMDP) model to maintain a probability distribution of where the robot is at all times, choosing actions based on that distribution [5, 7, 12]. The probability distribution is updated whenever the robot moves or turns (which tends to spread out the probability distribution) and whenever the robot thinks it perceives a landmark, such as a doorway or corridor junction (which tends to sharpen the distribution). Thus, while the robot usually never knows precisely where it is, it rarely gets lost.

The *path planning* layer determines efficient routes based on a topological map, augmented with rough metric information, and the capabilities of the robot. It uses a decision-theoretic approach to choose plans with high expected utility, taking sensor and actuator uncertainty into account [5]. For instance, if there is a reasonable chance that the robot will miss seeing a corridor intersection (and thus have to backtrack), the planner might choose a somewhat longer path that avoids the intersection altogether.

The navigation system is implemented as a collection of asynchronous processes, distributed over the three computers on-board Xavier, that are integrated and coordinated using the task control architecture (TCA). TCA provides facilities for interprocess communication (message passing), task decomposition, task synchronization, execution monitoring, exception handling, and resource management [9]. Using TCA, new processes can easily be added and removed from the system, even as it is running.

In addition to the layers described above, Xavier has processes that control the camera, provide speech generation, and monitor the robot's execution and recovery from failures [2]. For some experiments, a higher-level task sequencing layer is used to coordinates multiple, asynchronous tasks [4].

5.2.2   Interface Design

Since Xavier is intended to be an office-delivery robot, it needs a way for remote users (office workers) to interact with it. Our first interfaces, which were based on e-mail and Zephyr (a local, instantaneous messaging system), were somewhat stilted and had limited interactivity and flexibility. The World Wide Web interface was designed to make it easy for nonroboticists to interact with the robot. The

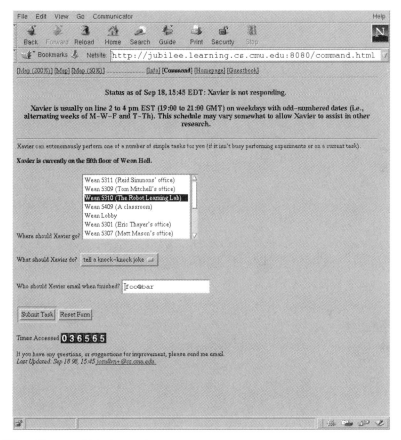

Figure 5.3:   Command interface web page.

command interface web page (figure 5.3) shows Xavier's current status (updated every 5 to 10 seconds) and provides a discrete list of destinations to send the robot (about a dozen different locations, mainly offices and classrooms, for each floor of our building), a list of simple tasks to perform at that location, and space to enter an (optional) e-mail address. The tasks that Xavier can perform at a destination include taking a picture with its color camera, saying "hello," and telling a robot-related knock-knock joke (the overwhelmingly favorite task).

When a user submits a task request, a confirmation web page is sent back immediately indicating when the robot will likely carry out the task (either immediately, if the robot is operational and not busy, at some time in the near future

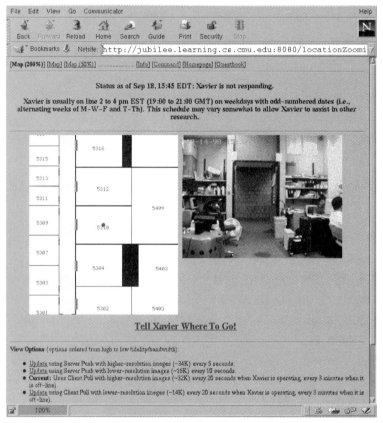

Figure 5.4:   Xavier monitoring web page.

if it is up and busy, or some time in the indefinite future if the robot is not currently on line). If the request includes a legitimate e-mail address, Xavier will send e-mail after it achieves the task, and it will include a mime-encoded image (gif format) showing what it saw when it reached that destination (plus the text of the knock-knock joke it told, if that was its task).

In addition to the command interface page, there is a monitoring web page that includes the robot's current status, a zoomable map of the floor Xavier is currently on, and a color picture of what it currently sees (figure 5.4). Both the map and the camera image are sent as gifs and are updated every 5 to 10 seconds. The map shows the area around the robot and its most likely pose, based on the probability distribution the robot maintains. Additional web pages linked to these command

and monitoring pages include information about the lab, statistics on Xavier's performance, a guestbook, and a "robot joke contest" page.

5.2.3   Robot Hardware

Xavier (see figure 5.1) is built on top of a commercial 24 inch diameter base, the B24 from Real World Interface. The base is a four-wheeled, synchro-drive mechanism that allows for independent control of the translational and rotational velocities. The base can move at up to one meter per second, although typically the maximum attainable speed is limited to around 60 cm/sec. Xavier has 1,400 amp-hours of batteries on board, and can operate continuously for two to three hours on a fully charged set.

The torso and superstructure of Xavier were designed and built by a class of computer science graduate students in 1993. The sensors on Xavier include bump panels, wheel encoders, a 24 element sonar ring, a Nomadics front-pointing laser light striper with a 30 degree field of view, and a Sony color camera mounted on a Directed Perception pan-tilt head. Xavier also has a speaker and a speech-to-text card. Currently work is being done to integrate speech recognition into the Xavier system.

Control, perception, and planning are carried out on two 200 MHz Pentium computers, running Linux. A 486 laptop, also running Linux, sits on top of the robot and provides for graphical display and communication to the outside world via a Wavelan wireless Ethernet system. The three on-board computers are connected to each other via thin-wire Ethernet.

5.2.4   Control

Since Xavier operates in a peopled environment, great care was taken to ensure that it would not harm (or unduly annoy) people in the building. Also, the robot is not always on line when users visit its web site, due to battery limits and other research demands for use of the robot. For these reasons, web users do not control Xavier directly, but instead request tasks that it (eventually) performs. User requests are queued and scheduled for execution at appropriate times (see below). If Xavier is on line when the request is made, the user receives an estimate of when the task will be handled; if the robot is off line, the request is queued. In this way, users can get access to the robot, eventually, even if they are unable to connect to it during

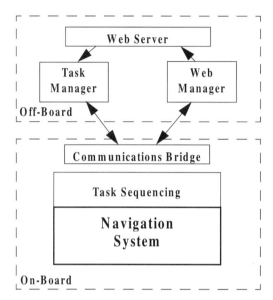

Figure 5.5:   Off-board and on-board system.

user operational hours (due to timezone differences, etc.).

Several additional on-board and off-board processes are used to control user interaction with Xavier (figure 5.5). The *task sequencing* layer is responsible for carrying out Xavier's tasks. This includes commanding the path planning layer to navigate to the requested goal location, centering the robot at the doorway if the destination is an office or classroom, and executing the given task (taking a picture, saying "hello," or telling a knock-knock joke). Currently, the task sequencing layer has only limited ability to monitor for task achievement or to recover from failures. Related work, however, has developed a much more sophisticated task sequencing layer [4, 11].

The *communications bridge* process, which also resides on board, is responsible for exchanging data between the on-board and off-board processes over the Wavelan radio link. Data exchange is via the TCA message passing facility, which is built on top of TCP-IP. The bridge process receives task and data requests from the off-board processes and forwards them to the appropriate on-board processes. In the other direction, the communications bridge receives position estimates, route plans, and camera images (gifs) from the on-board processes and forwards them to the off-board processes.

The rationale for having a single process responsible for all communication between on-board and off-board processes is that if wireless communication is lost for a while (which happens relatively frequently), the other on-board processes are not blocked trying to send data. In this way, Xavier autonomously (and safely) continues to carry out its tasks, even if it loses communication with the outside world.

The web site management system consists of two off-board processes, running on a Sparc5 workstation, that interface to a Netscape web server, also running on the Sparc machine. The *task manager* is responsible for queueing user requests, dispatching requests to the task-sequencing layer (via the communications bridge), and sending e-mail confirmations to users after each task is completed. The task manager uses a simple scheduling algorithm that tries to minimize the time until users' requests get executed. It computes the utility of going to a particular location as the sum of the utilities for each pending request for that destination, where the utility of an individual request is an exponential function of how long the request has been pending. The task manager then chooses the destination with the highest utility. Thus, it will be indifferent between a destination for which a single user has been waiting a fairly long time and one where many users have been waiting for shorter periods. Note that, in particular, there has been no effort to minimize the overall travel distance of the robot, since the original goal of the experiment was to stress-test the navigation system.

The *web manager* process is responsible for updating the web pages. It requests position information and camera images from the robot, creates a gif showing the robot in the map, and creates new HTML pages that include the robot status, current map, and camera images. It also creates new command interface pages, which differ depending on what floor Xavier is currently on. The web manager creates four types of pages that differ only in the bandwidth requirements needed for viewing on the Web. The pages with the lowest bandwidth requirements contain a low-resolution camera image and are updated every 20 seconds. The pages with the highest bandwidth requirements have a high-resolution image and are updated continually with streaming images.

5.3   The Setup

5.3.1   General Description

Xavier operates in a multistory building on the Carnegie Mellon campus that houses several departments, including computer science. The building has complete coverage by Wavelan base stations, so the robot can travel almost anywhere and still maintain contact with the outside world. Typically, Xavier operates on the fifth floor, where the lab is located, but occasionally the robot has been run on different floors. While work is being done to get Xavier to take the elevator autonomously, currently it is teleoperated to a new floor, its navigation software started with the map of the floor it is on, and it is told approximately where it is starting from. After that, the robot is on its own receiving task assignments from the task manager module, carrying them out autonomously, and feeding position information and camera images back to the web manager process.

After four years of traveling the corridors of the building, Xavier is seen as commonplace by most of the people in the environment. While some people try to "tease" the robot by blocking its path and some try hard to avoid it altogether, the vast majority of people are nonchalant about having a mobile robot in their midst. Certain concessions however, have had to be made to the inhabitants of the building. In particular, the only offices the robot visits are those where people have given explicit permission to use as destinations. The volume of talking (both amount and loudness) that Xavier does while traveling has also drastically been cut back. All in all, Xavier has tried to be a good "citizen."

The knock-knock jokes Xavier tells are especially popular (popular with web visitors, not so much with the occupants of our building). We created a "jokes contest" web page for people to submit knock-knock jokes that involve Xavier (example: "Knock knock" — "Who's there?"; "Xavier" — "Xavier who?"; "Zave-yer self from these awful jokes, turn me off"). New jokes continue to be submitted, even after four years, testifying to the collective creativity on the Web. Some visitors have even suggested allowing users to submit arbitrary messages for Xavier to say at its destination. Imagine the sociological consequences of that on the residents of the building!

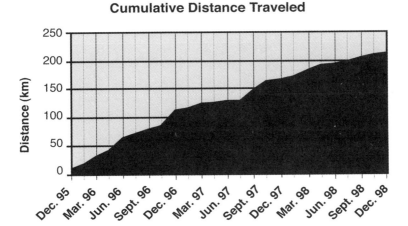

Figure 5.6:   Cumulative travel distance for Xavier.

5.3.2   Experience and Statistics

For the first several months of Xavier's operation, the robot ran for one to two hours almost every week day, usually in the late morning or early afternoon. Through the end of 1997, Xavier ran 2 to 3 times per week. It was also at this time that the robot was taken to different floors of the building. During 1998, the robot was online much less frequently, usually 5 to 10 times per month. Since then, Xavier is taken out infrequently, although the web interface and off-board processes are still up and running every day.

During three years of online operation (December 1995 through December 1998), Xavier received over 30,000 requests and carried out over 4,700 separate tasks (since requests are queued and then bundled together, the number of tasks is smaller than the total number of requests). In the process, Xavier operated for over 340 hours and traveled over 210 kilometers (figure 5.6). The average success rate for those three years in achieving tasks is about 95 percent, and that has increased to about 98 percent in recent months (figure 5.7, see also [11] for a discussion of the navigation results).

There are several explanations for the 2–5 percent failure rate. About half the failures were due to the wireless link being lost. While Xavier can successfully reach its destination without outside communication, the statistics are all kept by

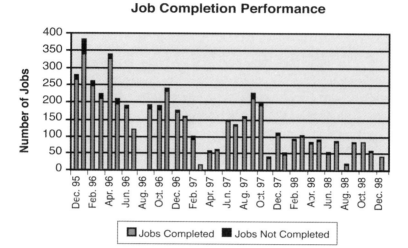

Figure 5.7:   Tasks attempted & successfully performed.

the off-board processes. Sometimes the wireless link goes down in a way that cannot be reestablished automatically and this is counted as a failure in the statistics. The remaining cases are true failures of the navigation system. In particular, the robot does not do well in crowded situations and in large open areas (such as the fifth floor foyer). On rare occasions, Xavier can get confused to the point that it cannot recover automatically. In such situations, manual intervention is needed. Fortunately, recent work in probabilistic navigation [1, 13] has overcome many of these problems.

While statistics on the robot's performance are important to the original goal of demonstrating reliable navigation, some perspective can also be offered on the users' experiences (gleaned largely through comments in Xavier's guestbook). First, it is apparent that the mobility of the robot has both positive and negative impacts on web interactions. The positive effect was that users felt that controlling a mobile robot remotely was a unique experience. Most of the effects of mobility on connecting with the Web however, are negative. Running on batteries limits the online time of the robot to a few hours per day. Thus, even though the web interface is always operational, most web visitors do not see Xavier in action when they happen to visit its web site. In addition, the need for wireless communication limits the bandwidth to the robot, which lessens the interactivity that can be achieved (however, this is becoming less of a problem as wireless communication improves).

Probably the most severe effect of mobility on online interaction is a sociological one: users can see what Xavier sees, but they cannot see Xavier itself. This is often disorienting, especially since the images are updated only every few seconds (higher bandwidth would definitely help). The Rhino tour guide robot [1] overcame this problem by using an overhead camera to track the robot, which was feasible in their case because they operated in an environment where the robot was usually in view of the single camera. Another approach would be to use a pair of robots, each watching the other. One of the visitors to the web site suggested putting a full-length mirror on one of the walls and letting people command Xavier to go there and look at itself.

On the other hand, the fact that Xavier is autonomous had mostly positive effects on online interactions. For one, autonomy mitigated the effects of low bandwidth and unreliable communication. Since the robot is being tasked at a high level (traveling to discrete locations), high bandwidth interaction is not strictly necessary. Even if communication is lost completely, Xavier can still continue achieving its current task. In particular, none of the navigation components are affected by loss of communication, so the robot's safety (and that of the people it encounters) is not affected. When communication is restored, the off-board processes reconnect with the on-board communications bridge, often automatically and usually without need to restart processes.

The only real negative impact of autonomy on online interaction is that commanding at a high level is not as interactive as teleoperation. Some users have expressed an interest in being able to choose an arbitrary location on the map for Xavier to go. Although the navigation system can handle that, for logistical reasons that level of control is not allowed. In particular, many occupants of the building are not too keen on having the robot visit them and tell them jokes on a regular basis.

One of the more surprising aspects of this experience is the degree to which people accept Xavier at face value. Given the nature of the Web, it would be comparatively simple to "fake" Xavier's travels with a series of canned images (much simpler, probably, than having to create a reliable, autonomous mobile robot). For the most part, however, few web visitors have ever questioned the authenticity of the robot. One exception occurred early on. Since Xavier uses a probabilistic navigation scheme, with a spatial resolution of one meter, it sometimes stops near, but not actually at, its destination. In such cases, the pictures e-mailed back to users would show walls, rather than doors or open offices. Occasionally, responses would come back questioning whether Xavier was really doing what it claimed. This was solved by training a neural net to recognize visually when the camera was pointed toward a doorway and then using a simple visual servoing routine to move the robot directly in front of the door. After this extension was implemented, comments about whether the robot was real were no longer received.

5.4   Conclusions and Future Work

Based on this experience, the authors have a number of observations that can guide the implementation of future online robots. The most important is the need for high-quality feedback. When the web interface to Xavier was first constructed in 1995, one priority was to minimize the bandwidth used so that the web interface would not interfere with other projects. The result is a rather slow refresh rate (5–10 seconds), which makes it difficult to see what Xavier is doing.

Since the original design, Xavier's computational power has tripled and standardized low bandwidth mechanisms and protocols such as Java and RealVideo have been developed and become ubiquitous. It is now possible, with a low computational overhead, to generate a continuous low-bandwidth, real-time video feed. Similarly, it is possible to construct dedicated Java applets so that map and position information can be displayed rapidly and efficiently (for instance, Minerva used such a mechanism effectively [13]).

An important part of the feedback mechanism is a guestbook where users can leave comments. This is invaluable, both for gauging the diversity and range of users and for soliciting suggestions. In addition to the usual collection of scatological and self-promotional comments, there are indications in Xavier's guestbook that an autonomous robot on the web strikes a particular chord with audiences not often associated with robots:

```
I am 4 and my name is Alexander. I am going to be 5 in
2 weeks and I want a robot for my birthday to clean up
my room and play pinch attack. My cat's name is Zoe. I
```

liked seeing pictures of you. I go to Brookview
Montessori and I do the hundred board. I would like to
play games with you on the computer. I have to go to bed
now, we are leaving now. Thank-you and goodbye. — Alex
T., March 7, 1998.

This is fantastic! I'm new to the web and feel like a
kid in a toy store for the first time. I happen to be 54
years old. — Mary H., October 9, 1998.

Taking this a step further, robotic web sites should host interactive "chat
rooms" for discussions related to robotics, technology and the web. Such
mechanisms have been used with great success by the Telegarden project [3].

Some users want more technical information about the robot, including details
on robot construction and programming. To accommodate this, a complete web
site was constructed around the original web interface. There are always
unanswered questions. While it is time-consuming, it is important to keep the
informational web pages accurate and current for disseminating technical
information to the public at large.

By far, the largest complaint is from users who miss those few hours when
Xavier is live on the Web (especially users in different time zones). We have tried
to alleviate this in several ways, including sending e-mail to notify users when the
tasks are completed. We are also considering notifying users a few minutes before
their queued requests are to be undertaken to give them a chance to see Xavier live.
None of this solves the fundamental problem that the Web demands immediate
feedback — continuous twenty-four hour presence is an important goal for future
online robots.

Overall, this online robot experiment has been successful. It has conclusively
demonstrated the reliability of the given navigation system, has given the robot
project good publicity, and has introduced many people around the world to the
wonders (and limitations) of autonomous mobile robots. The next step is to put the
robots to work. Work is now being done to put a pair of mobile robots in a new
building at Carnegie Mellon. These robots would cooperate to guide visitors,
answer questions about the department and university, deliver things (mail, faxes,
coffee) to inhabitants of the building and, of course, continue to tell jokes.

Acknowledgments

Xavier is the result of the collaborative efforts of many people. Special mentions go to Lonnie Chrisman, Joaquin Fernandez, Domingo Gallardo, Karen Zita Haigh, Nak Yong Ko, Yasushi Nakauchi, and Sebastian Thrun. Thanks also to the thousands of web visitors who have watched and commanded Xavier over the years.

References

[1] W. Burgard, A. Cremers, D. Fox, D. Hahnel, G. Lakemeyer, D. Schulz, W. Steiner, and S. Thrun. The Interactive Museum Tour-Guide Robot. In *Proc. National Conference on Artificial Intelligence*, pp. 11–18. Madison WI, 1998.

[2] J. L. Fernandez and R. Simmons. Robust Execution Monitoring for Navigation Plans. In *Proc. International Conference on Intelligent Robots and Systems (IROS)*. Victoria Canada, 1998.

[3] K. Goldberg, J. Santarromana, G. Bekey, S. Gentner, R. Morris, C. Sutter, and J. Wiegley. A Telerobotic Garden on the World Wide Web. *SPIE Robotics and Machine Perception Newsletter* 5:1, March 1996.

[4] K. Z. Haigh and M. M. Veloso. Interleaving Planning and Robot Execution for Asynchronous User Requests. *Autonomous Robots* 5:1, pp. 79–95, March 1998.

[5] S. Koenig, R. Goodwin, and R. Simmons. Robot Navigation with Markov Models: A Framework for Path Planning and Learning with Limited Computational Resources. In Dorst, van Lambalgen and Voorbraak, eds. *Reasoning with Uncertainty in Robotics,* vol. 1093 of *Lecture Notes in Artificial Intelligence,* pp. 322–327, Springer. 1996.

[6] S. Koenig and R. G. Simmons. Xavier: A Robot Navigation Architecture Based on Partially Observable Markov Decision Process Models. In D. Kortenkamp, R. Bonasso, R. Murphy, eds *Artificial Intelligence Based Mobile Robotics: Case Studies of Successful Robot Systems*, pp. 91–122. Cambridge: MIT Press, 1998.

[7] S. Koenig and R. G. Simmons. Unsupervised Learning of Probabilistic Models for Robot Navigation, In *Proceedings of the IEEE International Conference on Robotics and Automation.* pp. 2301–2308, Minneapolis MN, April 1996.

[8] N.Y. Ko and R. Simmons. The Lane-Curvature Method for Local Obstacle Avoidance. In *Proc. IEEE International Conference on Intelligent Robots and Systems (IROS)*. Victoria Canada, 1998.

[9] R. Simmons. Structured Control for Autonomous Robots. *IEEE Transactions on Robotics and Automation* 10:1, Feb. 1994.

[10] R. Simmons. The Curvature-Velocity Method for Local Obstacle Avoidance. In *Proc. of the IEEE International Conference on Robotics and Automation (ICRA)*, pp. 3375–3382, 1996.

[11] R. Simmons, R. Goodwin, K. Z. Haigh, S. Koenig and J. O'Sullivan. A Layered Architecture for Office Delivery Robots. In *Proc. Autonomous Agents '97*, pp. 245–252, Marina del Rey, CA, February 1997.

[12] R. Simmons and S. Koenig. Probabilistic Robot Navigation in Partially Observable Environments. In *Proc. of the International Joint Conference on Artificial Intelligence (IJCAI)*, pp. 1080–1087, Montreal Canada, 1995.

[13] S. Thrun, M. Bennewitz, W. Burgard, F. Dellaert, D. Fox, D. Haehnel, C. Rosenberg, N. Roy, J. Schulte and D. Schulz. MINERVA: A Second Generation Mobile Tour-Guide Robot. In *Proc. of the IEEE International Conference on Robotics and Automation*, March 1999.

6 KhepOnTheWeb: One Year of Access to a Mobile Robot on the Internet

Patrick Saucy and Francesco Mondada

6.1 Introduction

The research community has been used for years to remotely access, via standard communication networks, unique or expensive structures, such as supercomputers, important databases, unique information resources, or the World Wide Web (the Web), the e-mail box, or ftp servers full of software. With the growth of the Internet, one finds more and more devices such as coffee machines, cameras, telescopes, manipulators, or mobile robots connected to it. Despite the fact that one may spy on other people with hundreds of cameras, it is possible to interact only with a few robots which in addition sometimes have a restricted access [5].

Why is it so? There are several explanations. An installation with a robot is expensive. Regular maintenance is needed. Things are much easier with a camera because the interaction with the user is poor: one just sits and watches. Sometimes the user has the possibility of choosing a different orientation of the camera [16]. With a robot, one has a strong interaction. For instance, one can move on the floor and grasp objects with a mobile robot equipped with an arm [8]. Discovering the control interface, the user has to understand rapidly the goal of the site and what are the possibilities of the robot to achieve them. A famous example is the Mercury Project [1]. This kind of experiment is useful in the sense that it gives important information about the reactions of the users, the kind of equipment needed and the constraints of Internet. More information about devices connected to the net can be found in [9] and [10].

The long-term goal of the author's project is similar to those of the experiments described in the introduction, namely to provide access, through network communication facilities, to a complex and unique mobile robotics setup. The specific goal of this project is to provide a setup mainly to the scientific community carrying out research in mobile robotics control.

The goal of the project fits well with the activity of the lab. The Microprocessor and Interface Lab (LAMI) specializes in the development of mobile robots and tools for research in control algorithms. The remote experimentation is the logical extension of this activity.

This chapter describes the results of a first part of this project aimed at understanding the possibilities of the current technology, the reactions of users, and the requirements for the equipment of such a setup. The second part is briefly described in section 6.5. In comparison with other projects on the Web, presented in the introduction, our installation has some additional features:

- The controlled device is a mobile robot equipped with an on-board camera. This is also the case of some other setups that are partially available [6].

- Unlike other mobile robot setups, the robot here runs daily without the need of an external support.

- The interface has a live video feedback and runs in a standard web environment.

- Everyone has access to this robot. There is no distinction, such as registered user/guest, but only one person at a time can control the robot for a maximum of 5 minutes.

This chapter analyzes the net surfer's behavior through graphics and draws some conclusions about it. Section 6.2 describes the hardware and software components of this robot installation. Then an analysis of the log files of the Web server is discussed in section 6.2.4. The conclusions are presented in section 6.4. Finally, another concept of remote controlled robot on the Web is introduced in section 6.5.

6.2   Design

In this section the general design of the hardware and software architecture is presented, and in particular the interface design and the robot chosen for this experiment.

6.2.1   Hardware and Software Architecture

The host computer communicates with the robot via a RS232 link at the speed of 38,400 bits/s. The video signal is sent from the robot to the framegrabber in the PC in a differential mode on two additional wires.

An external camera (Canon VC-C1) is mounted on the ceiling above the maze

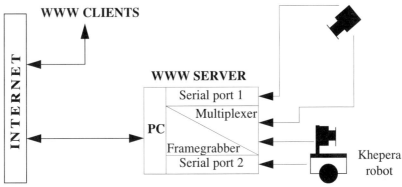

Figure 6.1:   Connections between the devices of the setup.

to give the user a global view. This is an interesting aid in planning a long displacement. A RS232 link between the PC and the camera allows the visitor to control the camera orientation and the zoom factor to have a better view of the current situation. The video is also wired to the framegrabber of the PC and multiplexed with the video signal coming from the robot camera. Therefore, it is possible to switch between the robot view and the external view. All these connections are summarized in figure 6.1.

The PC operating system is Windows 95. The Web server is the Personal Web Server from Microsoft. This server launches several Common Gateway Interface (CGI) scripts to perform the tasks. They communicate through shared named memory. Two programs run continuously:

• The first one takes the images and puts them in shared memory in JPEG format (160 x 120).

• The second one puts the IP address of every new user in shared memory in such a way that others CGI scripts can control user identity. A time-out variable is regularly decremented in memory. After 5 minutes, if the user has not disconnected, this program puts the user IP address in a list stopping control for during 30 seconds.

When a client performs an access to the control page, the server starts a program that continuously sends the images stored in the memory. This image feedback is based on the server push technique supported by Netscape but not by Internet Explorer at the time of writing. Others techniques can be used but have not been tested in this experiment [15].

On the client site, the user has access to a control page built mainly in plain HTML with clickable images. Three types of commands are available:

- Commands to control the robot movements (speed/position). The click coordinates are sent to the server, where a CGI script decodes and builds the corresponding order for the robot. The orders are sent to Khepera via the RS232 serial link.

- Commands to control the external camera movements (orientation/zoom). Here, too, the orders are sent to the camera via the RS232 serial link.

- Commands to switch the camera. A CGI script on the server acts on the multiplexer present at the input of the framegrabber.

A Java applet running on the client side regularly sends requests for information about the state of the robot and the time left to the user. A CGI script on the server answers these requests, collecting the information from the robot and the shared memory.

There is no local intelligence on the robot such as obstacle avoidance. This type of mechanism is not necessary because of the light weight of the robot, therefore there is no risk to destroy a wall or the robot itself. The advantage of having direct control is that the user can see the result of one's own action without any external contribution. The drawback is that the control of the robot is more difficult without help and under important delays.

6.2.2   Interface Design

The interface at the level of the client includes all possible operations that can be made on the robot, the external camera, and the multiplexer. The complete window available under Netscape is shown in figure 6.2.

The interface is composed of three columns, corresponding to the three types of commands that the user can perform:

- The left column contains controls for the orientation (pan and tilt) and the zoom factor of the external camera. The values of the parameters are given by clicking on the graduations.

- The middle column has the visual video feedback, the switch control of the two cameras, and the display of the robot status.

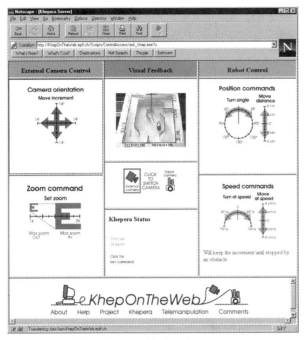

Figure 6.2:   The remote control interface provided under Netscape.

• The right column has two different control panels that allow one to send orders
  to the robot. The upper panel provides displacement commands, and the lower
  one gives speed commands. Each of these control panels has two parts. One for
  the rotation and one for the straight movement of the robot.

This interface is not the first one tried. Starting in 1995, a control interface was
developed based on a videoconferencing software (CuSeeMe). The results were
interesting but the control software was running only on MacIntosh, and therefore
it was moved to a pure web interface. The constraint is sending complete images
to the client while maintaining good performances. It is also possible to use
animated GIF and to send only the part that changes in the images, but a "caterpillar
effect" [15] can be observed.

The image size is small, so to give more access to the user, it is possible to display
the image larger than it is in the browser without loss of bandwidth.

Finally, more time has been spent on the design of the interface than on all other
engineering parts.

Figure 6.3:   The miniature robot Khepera with its on-board camera.

6.2.3   Robot Hardware

Khepera is a small cylindrical robot 55 mm in diameter and of variable height. Its small size allows for a suspended cable for power supply and other signals without disturbing its movements. A camera observing the environment can be placed without the need for wide-angle lenses. In its configuration shown in figure 6.3, Khepera is made up of three layers corresponding to:

- Sensory-motor board equipped with eight infrared proximity sensors and two motors
- CPU board equipped with a Motorola 68331 microcontroller
- Video board carrying a color CCD camera with 500 x 582 pixels

6.2.4   The Setup

The setup consists of a Khepera mobile robot accessible via the Internet and moving in a wooden maze, (figure 6.4). The labyrinth is constructed in such a way that the visitor can see various effects. The walls are generally higher than the robot, so that the visitor has to move around to explore the maze. Only some center walls are lower than the others. On the right there is a mirror that allows the user to watch the robot being controlled. The bottom left of figure 6.4 shows a ramp giving

Figure 6.4:   Khepera with its on-board video camera watching around in the maze, whose size is 65 x 90 cm. (Photo: H. R. Bramaz)

the robot access to a view outside the maze.

In this experiment the mobile robot Khepera was used; it was developed in the lab in collaboration with K-Team, the company producing and selling it. Khepera is equipped with an on-board CCD camera and is connected to a PC via a wired link. The camera sends the video signal to a framegrabber also placed in the same PC. This computer is also our web server and it is therefore connected to Internet. The user can access this installation via the web browser Netscape. Internet Explorer is not yet supported.

A virtual model of the setup has also been developed using VRML and Java. In this way the user can practice by running a simulation locally on a computer and without time delays constraints. A complete description of this implementation can be found in [3].

6.3   Experience and Statistics

The setup is still on line, but only the period between May 1997 and May 1998 was considered for this analysis. As explained above, the access to the virtual setup is not taken into account. All statistics are based on the IP addresses, therefore several persons from the same site will appear as a single user. Moreover, some Internet addresses are allocated dynamically and the same person visiting on separate occasions may appear as another person. In this paragraph, IP addresses, users, machines, and visitors have the same meaning. Next three terms are defined: *action*, *session*, and *nip*.

Action is a script run by the user to control the robot or the camera. A *forbidden action* is an action launched by a user who has not asked for the control of the robot from the presentation page. The user is not registered by the system and the action is refused. A *0 action* is due to a visitor who does nothing after having loaded the control page. If the control page was not loaded, we do not count it in *0 action* but in *no control page*.

A *session* or visit is defined as an uninterrupted access by the same machine with a maximal break of 10 minutes between two accesses.

Nip gathers IP addresses that could not be found in a domain name server (DNS).

6.3.1   General Statistics

Based on theses definitions, 27,498 visits were performed by 18,408 unique machines. Only 3,178 machines did more than one access. Their average return time is about twenty-three days with a typical delay between two actions of 13.6 seconds.

The most active session was issued from the United Kingdom. The user performed 630 actions distributed as follows:

- 344 actions for the control of the robot

- 243 actions for the control of the camera

- 43 actions on the switch of the panoramic and the embedded camera

This particular visit had a duration of one hour and ten minutes. This corresponds to an average of one action every 7 seconds. This visitor came back

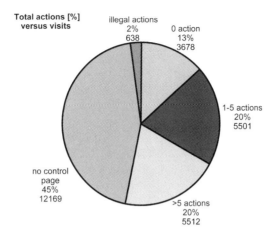

Figure 6.5:   Distribution of actions versus visits.

thirteen times but never completed so many actions again.

6.3.2   Actions Distribution

All other sessions were not as active as the one discussed above. Figure 6.5 depicts the usage of the site.

45 percent of users visited only the welcome page at our site. They did not request the control of the robot. There are two possible explanations for this:

- Browser type. Microsoft Internet Explorer does not allow visual feedback using the server push technique. The welcome page contains a warning about this problem. Statistics in [11] show that the distribution between the two main browser is: 50 percent Netscape, 45 percent Internet Explorer and 5 percent others.

- Type of access. The user can control the real or the virtual robot. The welcome page of the virtual setup was loaded by 11,283 machines.

20 percent of the users had access to the control page but did not perform any actions. The reasons for that could be:

- The robot was already controlled by other people.

- The user could not understand the goal of the site.

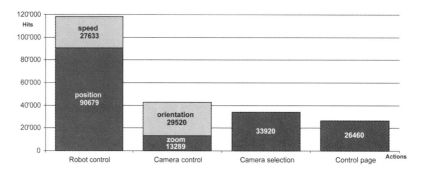

Figure 6.6:   Actions to control the robot and the camera.

- The image rate was too low (too big delays) or zero (browser type).

- There were problems while loading the clickable maps or the Java applet.

The previous graphic (figure 6.5) represents the actions distribution in general. Approximately the same distribution was found in other analyses not illustrated here:

- Action distribution versus months

- Action distribution versus number of visits

- Action distribution versus domain of the machines

- Action distribution versus internal lab use (demonstrations and verifications)

Figure 6.6 focuses on the effective use of the site represented by the actions. The robot is mainly controlled in position. This emphasizes the "wait and see" strategy developed by the users. Although the goal of the site is to control the robot, the camera is strongly solicited. Generally, a camera view is requested every second robot action. The visitor needs different view points to correctly understand the robot's location before deciding the next move. The importance of the column "camera selection" also shows the need to have both a panoramic camera and a camera on the robot.

The relationship between delays on the network and the number of actions performed is emphasized in figure 6.7. Only domains with more than 100 users who performed more than five actions are taken into account. There is a clear

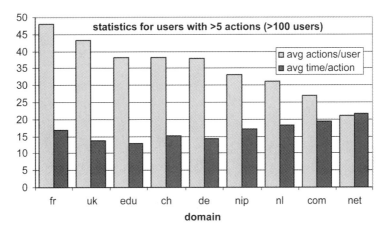

Figure 6.7:   *Relationship between delays and actions.*

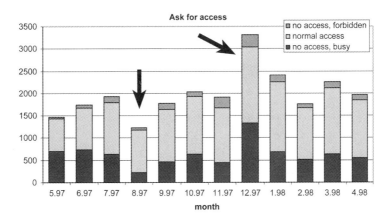

Figure 6.8:   *Accesses during twelve months.*

relation between the time/action and the number of actions/user.

Figure 6.8 depicts the influence of external causes on the evolution of the number of accesses along the twelve months considered.

The two months pointed to by the arrows are interesting. August 1997 shows a minimum due to an holiday period. August 1998 (not illustrated here) shows the same singularity. December shows a maximum. On December 8 the site was selected "Cool Robot of the Week" by NASA [12]. During the following month,

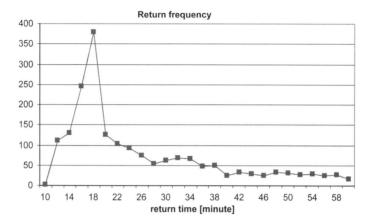

Figure 6.9:   Return frequency.

the number of accesses decreased and stabilized around 2,000 per month. People visit mainly once and do not return or come back for a second and last time. This behavior is most likely based on the "surf effect" of the Web. This result is verified by the statistics of the Australian telerobot [13].

6.3.3   Returning Visitors

Only 3,178 machines came back to this site. They made 12,268 of the 27,498 visits. Different analyses about the returns are shown in the next two graphics. Although return average is about twenty-three days, figure 6.9 exhibits an important peak of returns centered on 18 minutes.

1,096 of a 9,090 total returns are made after 12 to 22 minutes. This confirms the "surf effect." This peak is mainly due to people who come a second time to the site because they could not control the robot the first time (figure 6.8), so they retry later. There is no return below 10 minutes because two hits from the same machine made in an interval of 10 minutes belong to the same session by definition.

How do the returns influence the user behavior? Figure 6.10 attempts to answer this question.

This figure describes the distribution of the actions for each visit for four user categories performing a different total number of visits. The analysis is restricted to the first four visits to maintain a representative sample. Users who come back

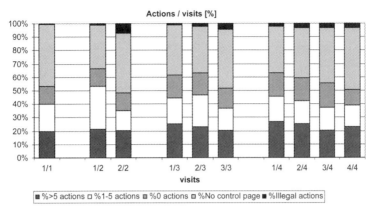

Figure 6.10:   User behavior versus visits number.

have acted during their first visit much more than the ones who visited only once. It could be expected that during their second visit users take advantage of their knowledge to do more actions. This is not the case. The number "of more than five actions" decreases slightly. People who come back appear to only look at the presentation page. The reason could be that they only want to show the site to some other people, get information, or simply verify that the server still exists. The number of illegal actions increases as well. It is possible that between two visits users have left their browser open on the control page. In this case there is a time out that makes the following action illegal. Most of the people come back after a short time, (figure 9).

6.3.4   Geographic Location Influence

All the previous figures are established on the basis of the accesses of 18,408 unique machines. This is less than the 60,000 hosts that visited the Mercury Project site in six months [1]. One of their graphics shows that 58 percent of the machines were located in North America and only 14 percent in Europe. In this case, the situation is the opposite with only 7 percent of the machines located in North America and 37 percent in Europe[1]. This could be because the Internet was less well developed in Europe than in North America. Another explanation is that this

---

1.   *The domains com, nip, and net have no geographic meaning. 51% of the machines come from these domains.*

site runs differently. The Mercury Robot is controlled through the means of absolute coordinates and this one through relative coordinates. It is therefore more difficult and time consuming to guide this robot from one point to another.

Switzerland and its neighbors (France, Germany, United Kingdom) were responsible for most of the actions. The geographic proximity (reduced access time) favors the use of this site. In other words, this site is functionally less attractive for users from far away because of the unacceptable response time. The domains com, nip, and net performed most accesses to this site but they made fewer returns to it than Switzerland and its neighbors.

6.4   Conclusions

The experience accumulated in this first test shows some important results. Such a physical environment is technically feasible with commercial parts. Everything from the robot to the camera, from the server to the framegrabber, are standard commercial products, but there are still problems. One problem is to introduce a complex setup in the Web where the rule is more "click and wait" than "read and click." Another is obviously the delay that prevents people from having a good interaction and to take interest in the site. This is not the only reason why users do not come back: the site is frozen, so there is nothing new to see in a second visit. Once a user controlled it, there is no reason to come back. You have to catch the net surfer's attention the first time. It is interesting to observe that the behavior of the users is generally independent of the country or the time the site is available, which shows that there is a global and stable Internet culture.

But the analysis of the Internet user behavior is not simple. The graphics and the analysis of the previous paragraphs show well that a complete understanding is difficult to get. Only few of our analyses are included in this article. Significative graphics are difficult to calculate because of the huge quantity of data in the log files. These were analyzed with the lab's software due to the specificity of the analysis made. It took two days of nonstop processing to scan the log files and to get the data in a form that could be handled to create a graphic. Even with such an amount of data there were still categories with not enough representative samples. When a graphic exhibits a feature, it is rather difficult to explain it with the general knowledge about the Internet. Moreover, one has little feedback from the users themselves. Only one mail for 1,000 accesses was received and it contained little information.

The reliability of the setup is good but two main problems had to be faced. As said above the PC server is running Windows 95. This system is not stable enough to allow the machine to operate without being regularly reseted to prevent a crash. The Personal Web Server was not well adapted for this usage. Sometimes it froze in a given state, then it is impossible to access the welcome page. This kind of test has been useful for a new project called RobOnWeb [14]. The project team has an additional reason to use Linux as operating system and Apache as server.

The other problem is mainly mechanical. Sometimes the robot is prevented from moving by being against a wall. Khepera is very light and its two motors are powerful. Therefore, against a wall, it is possible that it rises a little and one of its wheels hardly keeps contact with the floor. On the ramp, with the weight of the cables or due to dust, wheels can have less adhesion, making it difficult to control the robot.

6.5   Future Work

The main goal of this project is both economic and scientific. On one hand, a mobile robotics setup is expensive and could be shared between several research laboratories to reduce expenses. On the other hand, comparisons between different approaches in the field of mobile robotics can only be made on robots placed in perfectly identical conditions. This means not only that the robot has to be the same but also that the environment has to be identical in every detail like lighting conditions or color and type of floor. It is therefore scientifically of great advantage to share a common environment for the understanding of the actual control approaches.

During 1998, another benchmark (TeleRoboLab) was built. It is reserved only to researchers and has been available since the end of 1998. It takes into account the results presented in this chapter but adds a scientific dimension to telerobotics.

The new TeleRoboLab site provides complete access to all functionalities of a much larger robot, Koala. The user is able to remotely control the robot or to download C code. The access has been restricted to scientific users who have been registered. The setup complexity can be higher because the users are familiar with computers and robotics, but there is still a need for a good interface to let users concentrate on the algorithm without having to understand site functionalities.

Currently, there is no tool available to cope with delays on the Internet (perhaps RSVP in the future?) but this network is widely accessible. To improve real time access, the idea of this site is to be still accessible from the Internet but also from another network (ISDN). ISDN has a poorer performance than other networks like ATM but it is less expensive and easier to install. The environment of the robot (lights, sliding door) can be controlled through a web interface running Java applets.

At the time of this writing, the concept presented in this Section has been validated with a cross-platform navigation experiment. A neural controller evolved for the Khepera robot has been tested using the TeleRoboLab setup to analyze the adaptation capabilities of the control algorithm. This experiment was carried out between the LAMI and Mantra Labs located at the EPFL [7].

Acknowledgments

We would like to thank Edo Franzi and André Guignard for their important work in the design of Khepera. We also thank Jean-Daniel Nicoud and all the LAMI assistants who have tested our interfaces for their useful comments.

References

[1]   K. Goldberg et al. Desktop Teleoperation via the World Wide Web. *IEEE International Conference on Robotics and Automation*. Nagoya, Japan, 1995.

[2]   K. Taylor and   J. Trevelyan. Australia's Telerobot on the Web. *26th International Symposium on Industrial Robots*, Singapore, October 1995.

[3]   O. Michel, P. Saucy, and F. Mondada. KhepOnTheWeb: An Experimental Demonstrator in Telerobotics and Virtual Reality. *Virtual Reality and Multimedia Conference*. IEEE Computer Society Press, Switzerland, Sept. 10–12, 1997.

[4]   W. Burgard et al. The Interactive Museum Tour-Guide Robot. In Proc. of the *15th National Conference on Artificial Intelligence*, 1998.

[5]   R, Simmons. Where in the world is Xavier, the robot? Robotics and Machine Perception, special issue: *Networked Robotics*, vol. 5, No. 1, pp. 5–9, March 1996.

[6]   J. A. Fryer. Remote-control experiment using a networked robot. *Robotics and Machine Perception,* 5, Special issue: *Networked Robotics* 1:12, 1996.

[7]   J. Urzelai, D. Floreano. "Evolution of adaptive synapses: Robots with Fast Adaptive Behavior in New Environments," *Evolutionary Computation*, 1999.

[8]   http://freedom.artm.ulst.ac.uk/~antonh/research/Lynxmotion.html

[9]   http://www.yahoo.com/Computers_and_Internet/Internet/Entertainment/
       Interesting_Devices_Connected_to_the_Net/Robots/

[10]   http://queue.ieor.berkeley.edu/~goldberg/art/telerobotics-links.html

[11]   http://www.cen.uiuc.edu/bstats/latest.html

[12]   http://ranier.oact.hq.nasa.gov/telerobotics_page/coolrobots97.html

[13]   http://telerobot.mech.uwa.edu.au/

[14]   http://diwww.epfl.ch/lami/team/michel/RobOnWeb/

[15]   http://rr-vs.informatik.uni-ulm.de/rr/

[16]   http://graco.unb.br/robwebcam.html

7   RobOnWeb: A Setup with Mobile Mini-Robots on the Web

Roland Siegwart, Patrick Balmer, Caroline Portal, Cedric Wannaz, Rémy Blank, and Gilles Caprari

7.1   Introduction

The Internet has evolved at an enormous speed. It connects millions of people all over the world, independent of race, religion, or culture. The Internet became a governing technology during the last period of the twentieth century with an enormous impact on global society, giving access to communication, data, pictures, videos, and real time images of distant environments. Real physical interaction, however, is hardly available at the moment.

The goal of the project here is to take the first steps in adding a new dimension to the Internet by combining network technology with the capabilities of mobile robots. This way, Internet users can discover and physically interact with places far away from home.

7.1.1   Motivation and Related Work

Since the first robots appeared on the Internet in 1994 [1, 2], an enormous effort has been made by hundreds of research labs to push this technology. Most of the first robots on the Web, however, have been industrial robots operating within a structured and limited environment. Similarly, the first *mobile* robots on the Web have been mainly teleoperated systems [3, 4, 11] with very limited or even without autonomy working in highly structured environments.

To get real unbound interaction with a distant environment, a mobile platform is most desirable. It has no workspace limitations and thus allows for movement to places of interest and for real exploration of distant locations. Possible applications include visits to labs [6] or museums [4, 5], visit exciting cities or scenic natural parks, and exploring the desert, the poles, or even other planets [7].

Mobile robot systems connected to the Web or to any other network face three major problems:

- The network (Web) can introduce long time delays for which no upper bound can be guaranteed [7].

- The network enables unexperienced people without any sense for technology to guide the robots.

- The web interface has to be easy to use but with enough features to attract as many people as possible.

As a consequence of these issues, one can list the crucial points and the resulting specifications necessary for mobile robots on the Web.

- The system should never be harmed or crash due to long time delays.
  – The robot system requires a high degree of autonomy and a robust installation

- Internet customers should also be served appropriately during heavy network traffic.
  – A minimal data transfer should always indicate the immediate status, events, and robots position, for example, update of the robots' current position on an environment map would require only the transmission of three coordinates (x, y, θ).

- The update rate of the transmitted video images should be as fluent as possible to provide a good feeling for reality.
  – Data compression, optimal choice of resolution, focus the video feedback on areas of interest, for example, display only the part of a full frame image where the robot is.

- The web interface should be designed for "connect and play." Large introduction pages will scare away most of the customers.
  – For complex environments a training phase on a virtual reality simulation [3] of the same setup might help.

- The control strategy of the robot should be as intuitive as possible, for example, follow the doorway until the second intersection.

Keeping in mind these crucial points for online robotics, in 1997 an interdisciplinary project "*Mobile Robots on the Web*" [10] was started. Its main objective was to gain experience in this fascinating field and to establish the basics for Internet robotics. The main research focus were:

- Interactive web interfaces for mobile robots

- Modular framework for web robotics

This chapter presents an installation consisting of multiple mobile mini-robots on

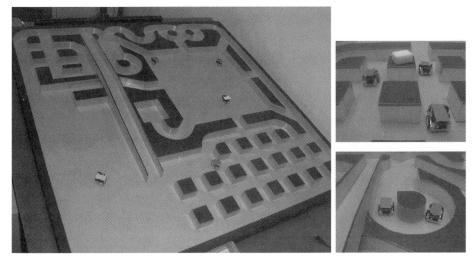

Figure 7.1:   The citylike maze with the five mini-robots Alice.

the Web and discusses the experience gained. It appears to be the first setup with multiple mobile robots, allowing up to five users to play and interact within the environment and the other users.

7.2   System Description

7.2.1   Setup and Robot Hardware

During the project "Mobile Robots on the Web," the final setup was developed and tested in different steps. This enabled the project to gain experience from untrained users right from the beginning and to optimize the interface and control concept.

The final setup consists of a citylike maze about 80 x 80 cm with five mobile mini-robots named Alice in it (figure 7.1). Each robot, identified by a different color patch, can be controlled by a user through the Internet. The robot's position is determined through the image of a camera mounted on top of the maze. In addition to the top view, a second camera allows the user to get a side view of the maze. To limit the data flow, only the region around the robot in use is submitted from the top view image to the individual user.

To control the robot movements, the user can select points of interest on the map of the maze. The requested new position is then processed by the navigation

Figure 7.2:   *The sugar cube robot Alice, the smallest mobile robot with real autonomy. (Photo: H. R. Bramaz)*

software (section 7.2.4), which generates segments of movements executable by the robots.

Robot Hardware

Alice (figure 7.2) is a mini-mobile robot whose size is only about 2cm^3 [8]. It is differentially driven by two watch motors with on-board sensors and a PIC μ-controller. Low-level algorithms running on the μ-controller allow for following walls, obstacle avoidance, and the detection of intersections. Alice has an energy autonomy of a full day, can communicate through an IR or radio link, and is inexpensive. It is therefore an excellent robot for this purpose, enabling the project to have multiple robots with high energetic autonomy on a small surface.

7.2.2   User Interface

The web interface (figure 7.3) consists of five modules for customer service. Each of these includes a server-side program and a client-side applet. They are completely independent and written in 100 percent pure Java.

- The *grab service* allows feedback from different video cameras. Users can select images from different sources.

- The *chat service* enables the users to send written messages to each other, either private or broadcast.

- The *lab service* draws an icon representing each robot at specific coordinates and angle. This allows updating the robot's position on an environment map and selecting new goals within the map.

- The *command service* allows the user to send commands to a socket. These commands are then handled by the C server described in the next section.

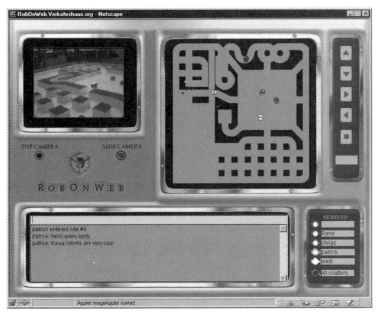

Figure 7.3:   User interface of the setup.

- The *login service* enables a new user to connect to the site and handles the nickname of the users as well as their IP address.

The maximum simultaneous connections tolerated for each service can be specified independently (for example, one person controls a robot with video feedback, and four other persons have video feedback only). In the standard installation, however, each robot will be controlled by only one user.

The graphical interface seen by the user is shown in figure 7.3. On the top left, the user sees the video image, either from the top camera or from the side camera. The current robot position is always indicated in the environment map on the top center. The same map is also used to specify the new robot position. This is done by simply clicking with the mouse at the requested position within the map. On the top left one can see the buttons for direct robot control. They allow to specify the robots movement (forward, backward, turn left, turn right, stop). Depending on the data transfer rate, however, direct control will have almost no value. Therefore the robots are usually controlled by selecting the goal position in the map.

The bottom part of the interface is used for the chat service. It enables the users

Figure 7.4:   Setup at computer exhibition, Lausanne 1999.

to communicate with each other in the setup. It might be the most appealing part of the setup, because it stimulates collaboration of the users.

7.2.3   The Installations

Installation at the Computer Exhibition '99, Lausanne

First testing of this setup was done on a temporary installation at the computing exhibition Computer '99 in Lausanne, Switzerland (figure 7.4). The three Alice robots of the setup were only accessible through the local network at the exhibition. The aim was to gain first experience with our multirobot system from public users. The feedback of around 500 local users was positive and lead to the conclusions that a multirobot setup is attractive and people are able to control the robots without prior explanations. Moreover, it was recognized that many users started to interact with the other robots and used the chat line to initiate special interactions. On the other hand, a lot of minor problems were experienced with the robot hardware (e.g., on board sensors) and the sensitivity of the vision system to environmental light.

Figure 7.5:   Permanent installation of the mini-robot setup at the Museum of Transportation in Lucerne, Switzerland.

This experience triggered many improvements on the hardware and illumination of the setup.

Permanent Installation at EPFL

Shortly after Computer '99, we opened the web site *http://RobOnWeb.epfl.ch* as a permanent demonstrator in the lab in Lausanne (Switzerland). Compared to the Computer '99 setup, this installation now uses five Alice robots and is open to everyone connected to the Internet with a Netscape browser and the required Java plugin.

   The aim of the setup was to verify the robustness of the software and hardware before final installation at the Museum of Transportation in Lucerne. It is still in operation and will remain online for at least another year.

The Verkehrshaus Setup

Finally, after acquiring the feedback from the two preliminary installations, the final setup became operational on 1 December 1999 at the Museum of Transportation in Lucerne, Switzerland (figure 7.5). Through the permanent address *http://RoboOnWeb.Verkehrshaus.org* on-site visitors as well as the whole

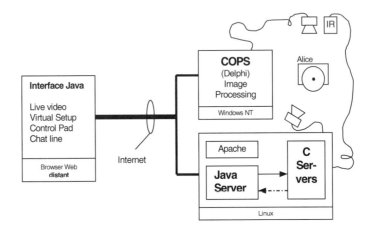

Figure 7.6:   Schematic overview of the robot control and interface framework.

Internet community could control the five robots during the opening hours of the museum.

7.2.4   Software Architecture and Design

One of the primary goals of the project was to develop a modular software architecture and programming environment able to generate new processes and to connect them easily. This facilitates and accelerates the creation of complex systems and architectures. Changing from one robot type to the other is therefore possible in a "plug and play" manner by only changing the definition file of the robot. The approach also allows different robots to run in parallel within the same environment.

Software Elements

The software is built out of three parts (figure 7.6): the image processing COPS (Color Object Position Server), developed in Delphi and running on a Windows NT server, the C servers controlling the robots (navigation, energy, etc.) running on Linux, and the Java part running on Linux with the Java applets for the user interface and the Java servers managing the applets, the users and the communications with the C servers.

The COPS Image Processing

Each robot in the maze is recognized by the image-processing algorithm through its rectangular color patch (one color per robot). The colors used are: red, blue, green, yellow, and white. COPS is in charge of detecting these patches and finding their position/orientation in a coordinate system bound to the physical maze (the image distortion due to the lens is partially corrected by calibration). To do so, COPS groups directly connected pixels of the same color category and, after some filtering, are able to determine precisely the characteristics of the patches.

The configuration of COPS allows to set different parameters:

- A tolerance range for each color to be detected within a discrete color space of 16 bpp.

- The filtering algorithms and parameters to eliminate noise or parasitic potential patches due to light reflection.

- The complete configuration of the acquisition system (contrast, brightness)

- The bounds of the image processing, the coordinate system, and the scaling.

Moreover, COPS is an Internet server that shares the coordinates, the orientation and parts of the image with the C and Java servers.

The Applets and the Java Servers

The execution of orders given by the Internet users to the robot requires two stages. The first one is handled by the Java Servers and their applets (figure 7.7), the second one by the C servers.

The four applets on the user's side (bottom of figure 7.7) manage the login of the clients on the site, inform them of the state of the setup (number of robots, position of the robots, nickname of the connected users and which robot they are guiding), receive the user's control requests and send them to the servers.

The six servers (center of figure 7.7) communicate with the applets but also with the rest of the software (COPS, C processes). The main server called JavaManager launches at start-up all the servers according to the number and name specified in the initialization file. It manages the internal communication (between the Java servers) and the external communication (between the manager of the site and the servers). It is also in charge of checking the status of the processes that are launched.

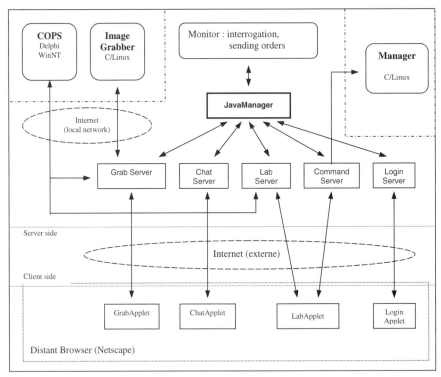

Figure 7.7:   Schematic of the Java server and its applets.

The LoginServer is the first to interact with a new user who connects to the site. The applet LoginApplet (in a direct communication with this server) sends the nickname specified by the user and makes a request for a connection. The LoginServer accepts the connection if the setup is able to welcome a new user (depending of the maximum number of connections configured in the initialization file) and assigns a robot to the user. It is also able to restrict the number of connection to some IP addresses (this can be parameterized in the initialization file). If the connection is accepted, a new page with the other three applets is loaded to the user's interface.

The applet ImageApplet is in charge of the video feedback. The user may choose between two different views: the upper view and the side view. In each case, the applet generates a request to the GrabServer to get the image; the GrabServer then sends the requested image. For the upper view, the GrabServer makes a request to COPS that replies with a reduced image showing only the area around the robot of the individual user. For the side view, the request is made to the C process ImageGrabber that handles the side- view images.

The applet ChatApplet allows the users to communicate among themselves through a chat line. The principle is simple: the applet sends the users' messages to the ChatServer that distributes them to the concerned applets.

Finally the applet LabApplet allows the user to know the position of the robots within the map of the maze and to control them. The LabServer continuously sends the position coordinates of the robots (given by COPS) to the applet, thus allowing it to update the map easily with the indicated robots' position. If the user clicks on the maze, the applet sends the selected goal position to the LabServer that routes it to the C processes as a high-level command (described in the next section). The user can also steer a robot with simple commands by clicking on the buttons (right, left, forward, backward arrows, and stop). These commands are first sent to the CommandServer that then sends them to the C processes.

This Java architecture needs robust internal and external communication among the processes, while respecting the modularity of the system.

Description of the C Servers

The C server controls the robots (navigation), grabs the image of the side camera, and is the communication hub between the Java servers and the robots (figure 7.8). The software architecture of the processes running on the C server is modular. The system is easily adaptable to different robots and environments through changing the definition files.

The architecture is based on a shared memory zone containing the parameters of the different processes, the topologic-metric model of the maze, and the description file of the robot. The shared memory can be viewed and changed by the Monitor Server process that allows one to evaluate and modify every parameter while the setup is running. This also simplifies the exchange of information between the different processes. The second core process of the architecture is the Manager. It launches the other processes at start-up and manages them during operation. Processes can be reset after important parameter changes and stopped or rebooted after version changes at run time. Having the parameters in the shared memory guarantees that the new process will always access the latest parameter set.

The intelligence for reliable control and navigation of the robots is distributed

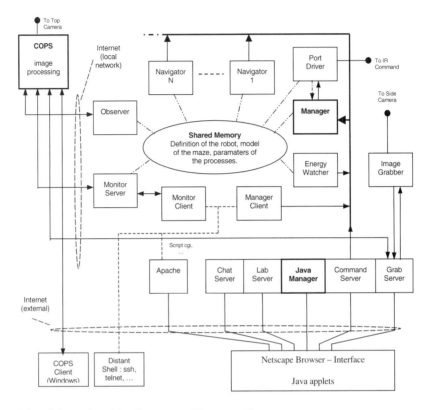

Figure 7.8:   Schematics of the C-server and its connections.

among the robot's on board microcontroller, the C server, COPS image processing system, and the Internet user. The control architecture considers the special issues of robots controlled through the Internet, which are the unknown time delay and the untrained users. This implies that the user has to be able simply to specify the robot task, without knowing any technical specification of the robot. In a similar way, the Java processes have to manage the user without knowledge of the type of robot. To do so, a standard high-level language was defined (used by the Java interface). A click on the maze stands for the *GoTo* command or a click on the right arrow for the *Right* command. A robot-specific definition file then translates the high-level commands to low-level actions. It is easy to translate the high-level command *Right* to the low-level equivalent available on almost every robot, but it might be much harder to find the equivalent of the *GoTo* command.

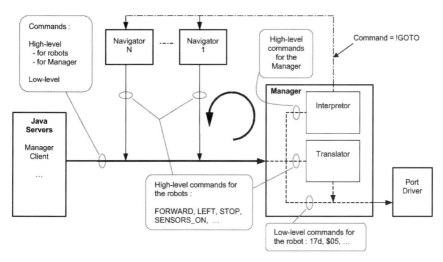

Figure 7.9:   Navigation and control.

The Manager handles proper translation of high-level to low-level commands based on the definition file of the robot. At this level the specific features of the robot appears. The definition file might contain either an easy equivalence (*Right* = 8d) or a more complex equivalent defined in a script language. A script is usually not powerful enough, however, for real-time navigation. Therefore special navigator algorithms (figure 7.9) were developed in C that handle commands such as *GoTo* much more efficiently. Consequently the Java part of the software and the user can forget about the low-level software.

When the user selects a new robot task through the interface (figure 7.3), a high-level request is sent to the C servers via the Java servers. The request is then sent either directly to the robot or to the Manager for interpretation (figure 7.9). In the first case, it will be translated with the definition file of the robot and sent via the Port Driver. In the second case — for example, a *GoTo* command — the Manager sends the command to the concerned Navigator for processing. The Navigator then generates a sequence of directly executable high-level commands based on the same definition file. This renders the Navigator modular and independent of the type of robot.

The Navigator contains several strategies that can be changed while they are running. The first strategy is based on a segmentation of the path into small steps determined by the model of the maze. Each step is executed through commands implemented in the robot. At the end of each step, the navigator verifies the robot position. This strategy results in a robot movement that is not very fluent because the robots stops after each step. The second strategy implemented in spring 2000 on the setup is quite different. It takes full advantage of the robot's basic behaviors. The robot is not controlled in little steps, but by starting a certain behavior. If the robot follows a straight path through the maze with various intersections, for example, the Navigator only starts the behavior *Straight,* and changes it to *FollowLeft* as soon as the robot has reached an intermediate node on the path. The robot's position is supervised through the position updates of COPS. The crucial point with this second strategy is a good geometrical definition of the maze and a reasonable position update. Both are satisfied within this setup.

7.2.5   Discussion

The primary goal within the described project was to develop a modular framework allowing for easy connections of all kinds of mobile robots to the Internet. In an early stage of the project, it was decided to use Java for all the elements concerned with the user interface. This choice might have the advantage of being future oriented. During the project, however, it was realized that Java is still too premature to be modular and machine independent. The project suffered from these problems related to Java, and as a result the accessibility of the setup is at the moment limited to Netscape on a PC. Additionally, downloading the required Java plug-ins scares off a lot of potential users, especially those connected through low bandwidth telephone lines.

Nevertheless, the installation at the Museum of Transportation in Lucerne is a great success. It is sponsored by the largest telecommunication company of Switzerland, Swisscom, with the intention of attracting peoples to use the Internet. The attractively displayed setup draws people's attention and has become the main attraction within the telecommunication area of the museum.

7.3   Performance

As mentioned in section 7.2.1, the final setup was tested through some preliminary setups at a computer exhibition and at the Swiss Federal Institute of Technology. These preliminary setups, however, were only experimental installations without representative statistics. Therefore only the operational data and experience of the final setup is discussed in the following sections. It is in operation since 1 December, 1999 and will be accessible at least until the end of 2001.

7.3.1 Operational Experience and Statistics

As mentioned earlier, the setup at the Museum of Transportation is online during the hours of the museum which are around 9 hours per day for 365 days a year. Since it went online on 1 December, 1999, the system has been working without any major problems. During the first three months only three crashes of the server were encountered, which were resolved be rebooting the machines. The setup requires minimal support for changing the batteries of the robots. Four sets of five robots are always ready at the site. None of them has shown any degradation or malfunctions up to now.

One robot Alice is displayed in front of the setup to represent the robot technology used. Even though it was well protected, it got stolen and had to be replaced. This might demonstrate that Alice gets a high-level of attention.

7.3.2 Statistics

This analysis considers the period between December 1999 and February 2000. Statistics are based on IP addresses. Because some systems allocate the IP addresses dynamically, it might happen that one person visiting the site several times appears as different people. Moreover, different people from the same site will appear as only one.

Advertisement for the site was carried on Swiss TV on 2 December 1999. Therefore most of the visitors are from within Switzerland (figure 7.10). No advertisement were made outside Switzerland.

General Statistics

During this period from December 1999 through February 2000, 2,135 sessions were performed by seventy-three different machines (figure 7.10). Note that at this point a lot of sessions were "test sessions" performed to control the setup (Verkehrshaus.org). The visitors came mainly from Switzerland, but they also came from overseas and commercial domain (.com).

Figure 7.11 presents the daily traffic. There is a first peak on 4 December. This

Origins of the users

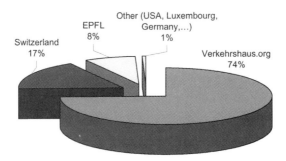

Figure 7.10: Percentage of sessions ordered by addresses.

Number of connections per day

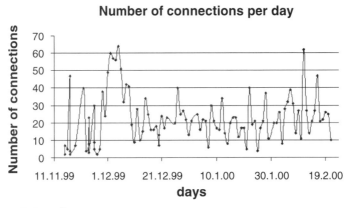

Figure 7.11: Daily traffic.

is due to the inauguration of the setup at Verkehrshaus 2 December and the related publicity. On 3 December, the Swiss TV broadcasted a report on the setup and thus stimulated a lot of visits. On average around 25 visits per day were encountered, with small peaks during the weekend when people have more time to surf.

Accessibility of the Setup

As mentioned earlier, one needs a recent version of Netscape (4.6.1 or higher) and the Java JRE plugin 1.2.2 to access the setup. The first page of the site tests the configuration of the user and offers the possibility to download the missing component(s). Netscape and the plug-ins, however, have a total size of 20 Mb.

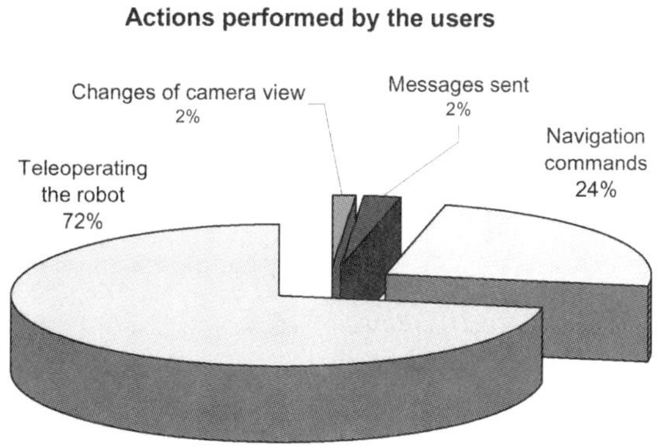

Actions performed by the users

Figure 7.12: Percentage of the different actions selected by the users.

Of the 2,135 sessions, one-third were directed to the install page because they did not have the right configuration. In 700 sessions only 280 persons downloaded the Java plugin (5 Mb) and 80 Netscape (15 Mb). Only 4.5 percentage of the surfers with a wrong configuration took the time to instal the right one; this is mainly a question of time. First the software has to be downloaded (with a 33.6 modem, it can take a minimum of 1 hour 20 minutes), then one has to install it and restart the computer. Visitors do not want to wait to play with the robots. If the plug-ins had already been integrated in Netscape, or if the Java virtual machine was dissociated from the browser, the number of visits would have been significantly higher.

Discussion

Once the site is entered, an average visitor session takes around 5 minutes. During this time the user has several possible options: guide the robots (direct teleoperation), navigate the robots (selecting the new goal on the map), chat with the other users, or change the camera view.

As seen in figure 7.12 most time was spent on guiding the robot directly by using the forward, backward, and turn buttons. This might be the most obvious action on the interface. However, if you consider the distance covered, the number of navigation actions results in much higher value. This because each navigation action automatically combines in average around ten to fifteen guiding actions.

About one fourth of the actions were controlling the robot be clicking on the map. Unfortunately the chat tool was not used very often. Perhaps some sort of game that needs collaboration of the robots should be added to the setup. This might increase the use of the chat line and stimulate interaction between the different users of the setup.

7.4   Conclusions and Outlook

A modular framework for the control of small mobile robots connected to the Internet has been presented. It enables the easy exchange of robots and environments. A permanent setup of five small mobile robots has been permanently installed and everyday is fascinating many people at the Museum of Transportation and all over the world. The main interface software was written in pure Java. This might be an opening for the future, but it restricts the access to the setup momentarily to Netscape browsers on PCs. The statistical data show an average of twenty-five users daily. Due to the restricted opening hours of the setup, the majority of the external users are from Europe.

It appears that all of today's mobile robots on the Web are operated within a structured, well-known environment. Even if some of the mobile robot systems are working in populated areas (universities, museums, exhibitions), localization, path planning, and navigation are drastically simplified due to the precise knowledge of the environment. One of the great challenges for the future is to connect mobile robots to the Web that are able to explore unknown environments. To do so, further advances in mobile robotics are necessary. Research and development continue toward this ambitious and fascinating goal and the authors are convinced that this dream will become a reality within a few years.

Acknowledgments

We would like to thank the Swiss Science Foundation, Swisscom, K-Team and the various scientific and industrial partners for supporting this project.

References

[1]   K. Goldberg et al. Desktop Teleoperation via the World Wide Web. *IEEE International Conference on Robotics and Autonmation.* Nagoya, Japan, 1995

[2]   K. Taylor and J. Trevelyan. Australia's Telerobot on the Web. *26th*

International Symposium on Industrial Robots. Singapore, October 1995

[3]   O. Michel, P. Saucy, and F. Mondada. KhepOnTheWeb: An Experimental Demonstrator in Telerobotics and Virtual Reality. *Virtual Reality and Multimedia Conference*. IEEE Computer Society Press. Switzerland, Sept. 10 – 12, 1997.

[4]   J. A. Fryer. Remote-control experiment using a networked robot. *Robotics and Machine Perception*, special issue: *Networked Robotics*, vol. 5, no. 1, p. 12, March 1996.

[5]   W. Burgard et al. The Interactive Museum Tour-Guide Robot. *Proc. of the 15th National Conference on Artificial Intelligence*, 1998

[6]   R. Simmons. Where in the world is Xavier, the robot? *Robotics and Machine Perception*, special sssue: *Networked Robotics*, vol. 5, no. 1, pp. 5 – 9, March 1996.

[7]   T. B. Sheridan. Space Teleoperation Through Time Delay: *Review and Prognosis. IEEE trans. Robotics and Automation,* vol. 9, no. 5, pp. 592 – 606, 1993.

[8]   G. Caprari et al. The Autonomous Micro Robot ALICE: A platform for Scientific and Commercial Applications. *Proc. of MHS '98*. Nagoya, Japan, November 25 – 28, 1998

[9]   K. Arras and R. Siegwart. Feature Extraction and Scene Interpretation for Map-Based Navigation and Map Building. *Proc. SPIE's Conf., Intelligent Systems and Advanced Manufacturing*. Pittsburgh, October 13 – 17, 1997

[10]   Siegwart R., Wannaz C., Garcia P., and Blank R. Guiding Mobile Robots through the Web, *Workshop Proc. of IEEE/RSJ International Conference on Intelligent Robots and Systems*. Victoria, Canada, October 12 – 16, 1998

[11]   Siegwart R., Saucy P. Interacting Mobile Robots on the Web, *Workshop Proc. of 1999 IEEE International Conference on Robotics and Automation*. Detroit, May 10 – 15, 1999 (http://www.cs.berkeley.edu/~paulos/papers/icra99/)

8 The Performance of Mobile Robots Controlled through the Web

Hideki Hashimoto, Noriaki Ando, and Joo-Ho Lee

8.1 Introduction

During the past several years, robotic research using computer networks has shown enormous development. Now the Internet has spread widely around the world and it can be used inexpensively. Nowadays teleoperations also use computer networks like the Internet as communication media. This chapter is a study of mobile robot teleoperation via the Internet from 1997.

8.1.1 Motivation

To date, the authors developed four different teleoperation systems and experimental simulation systems:

- *System I*: Basic teleoperation system, which has original user interface and joystick type controller, was developed in 1997 [1].

- *System II*: Online mobile robot teleoperation system, where operator controls the mobile robot with web browser developed in 1997.

- *System III*: Time delay emulation system for evaluation usability of Internet-based teleoperation system developed in 1998 [2].

- *System IV*: Simulation system to evaluate operability of human in time-delayed teleoperation system developed in 1999 [3].

At first, a simple teleoperation system was developed for a mobile robot, which had an original user interface with UDP/IP (system I). A online mobile robot teleoperation system was also developed (system II). In this system, an operator controls a mobile robot with a web browser through the web page composed of Java/JavaScript and CGI. The system was developed to investigate network-connected robot acts as physical agents in network robotics. In these experiments it was believed that robots would be a physical agent as a new medium in the Internet. In these systems, however, some problems became clear. One of the problems is delay during the Internet communication. Another problem is controllability, which is caused by human interfaces. This is often the case; since the operator controls the mobile robot only with the narrow field of visual image, it is difficult to understand the present situation.

Perceiving speed is also difficult for the operator, because operation display, which shows images from a robot camera, does not provide sufficient information. As a controller, a joystick is easy to use but difficult to correct errors during path control. From experience with the system I and system II, an evaluation of human factors appeared warranted.

8.1.2 Related Works

When teleoperated robots are connected to a network, real-time operation and synchronization of data exchanges are serious and important problems. One of the solutions to these problems is improvement of the stability of teleoperation control systems. Mark W. Spong showed a method to compensate for the time-delays in communication using the scattering theory formalism [4]. He discussed the problem of time delays inducing instability in force, reflecting teleoperators and the scattering theory formalism that provide a methodology for time-delay compensation. Kevin Brady and Tzyh-Jong Tarn showed a method for controlling telerobots over vast distances, where communication propagation delays exist [5]. They presented a canonical state space formulation taking into account the time-varying nondeterministic nature of the control and observation delays, and they derived a model of the delay characteristics for the communication media.

Another solution is giving high intelligence to robots to control them by high-level commands, which is called supervisory control. Luo et al. proposed a system that shows remote images on a web browser that controls a mobile robot, carrying a robot manipulator with the browser [6]. Kawabata et al. in RIKEN reported on the role of intelligence of a robot when an operator controls the robot based on only images from the robot [7].

A large number of studies have been made of teleoperation. Most of the studies are focused on the stability of systems, intelligence of robots, the human interface, and so on. A teleoperation system can be represented by five subsystems: the human operator; the human interfaces (e.g., master devices, visual display, etc.); the communication media; the slave devices (e.g., actuators, sensors etc.); and the environment. The information that appears to be lacking is the property of human operator, of which modeling is a difficulty, in teleoperation. Therefore, an evaluation of the influence of time delay in teleoperation quantitatively was evaluated.

8.1.3   Contribution

This chapter shows a quantitative evaluation of operability that depends on communication time delay in teleoperation. In addition, it proposes guidelines of human interfaces for network robotics, including teleoperation. For this purpose, some mobile robot teleoperation systems were developed (system III and system IV) to investigate the influences of human operator in teleoperation with communication time delay.

The chapter's next section states the design of the teleoperation system: software architecture, interface design, hardware, and control section. Section three explains the system setup and experimental results. Section four discusses the experimental results and concludes this chapter.

8.2   System Description

This section introduces the teleoperation systems and experimental system.

8.2.1   Hardware

Figure 8.1 shows the mobile robot "Kani," which is commonly used in systemsI, II, and III. Kani has two driving wheels, which are controlled independently, and two caster wheels. This mobile robot has two microprocessor units: one is used as a web server for network handling processing and the other is used for motion control. These microprocessor units are connected to each other through serial (RS-232C) ports for sending data of direction, velocity, and other control parameters. Web server and other network connection programs are run on FreeBSD 2.0.5 on the microprocessor unit (Pentium 133MHz/32MB memory). This robot has a wireless LAN adapter, which has a 2Mbps bandwidth, and a CCD camera (SONY EVI-D30), which is able to control pan and tilt motion. Video images are transmitted by UHF video transmitter, and images are received by a video tuner board (Bt 848 based video capture and tuner board), which is installed on the remote site host PC.

In the mobile robot Kani, two driving wheels are controlled independently by

Figure 8.1:   Mobile robot Kani.

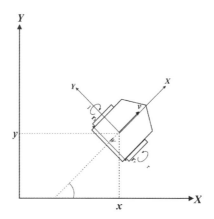

Figure 8.2:   Kinematics.

a velocity vector. The parameters of Kani are shown in figure 8.2. This mobile robot's moving model is described as follows,

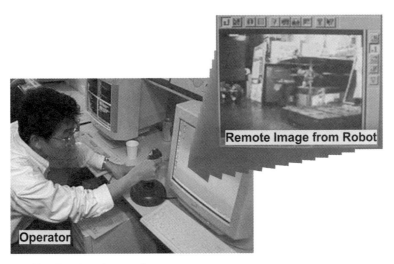

Figure 8.3:   User interface (system I).

$$
\begin{bmatrix} v \\ w \end{bmatrix} = \begin{bmatrix} \dfrac{r_r}{2} & \dfrac{r_l}{2} \\ \dfrac{r_r}{W} & \dfrac{r_l}{W} \end{bmatrix} \begin{bmatrix} \omega_r \\ \omega_l \end{bmatrix} \quad ; \quad \begin{bmatrix} \dot{x} \\ \dot{y} \\ \dot{\theta} \end{bmatrix} = \begin{bmatrix} \cos\theta & 0 \\ \sin\theta & 0 \\ 0 & 1 \end{bmatrix} \begin{bmatrix} v \\ w \end{bmatrix}
\tag{8.1}
$$

$$
x = \int \dot{x} dt \; ; \quad y = \int \dot{y} dt \; ; \quad \theta = \int \dot{\theta} dt
\tag{8.2}
$$

Where r_r , r_l are radius of wheels, W is width between wheels, ω_r , ω_l are angular velocity of wheels, v is translation velocity, w is angular velocity of the robot. The PID controller controls two motors, which drives wheels.

8.2.2   User Interface

This section describes the interface design.

First, systems I and II are discussed, which are the premise of the evaluation systems. In system I, a general teleconference system was used to get remote site images, shown in figure 8.3. An operator controls the mobile robot with a joystick while watching remote images. The remote image size is 320x240 pixels in the teleconference system. System II, shown in figure 8.4, is a online teleoperation

Figure 8.4:   User interface (system II).

system. A video transfer system was developed via http to watch remote images with a web browser. An operator controls the remote mobile robot with a web browser pushing buttons on the page. The size of images, shown on the web browser, is 320x240. The maximum frame rate is about 10fps.

The operability evaluation system (system III), shown in figure 8.5, used a joystick for user interface. In the teleoperation delay emulation system, remote images are transmitted by video transmitter and received by video capture board with a TV tuner. Received images are saved to FIFO, and these images are shown after a set delay time. This system can control delay time from 0 second to about 7 second, and it can also control frame rate.

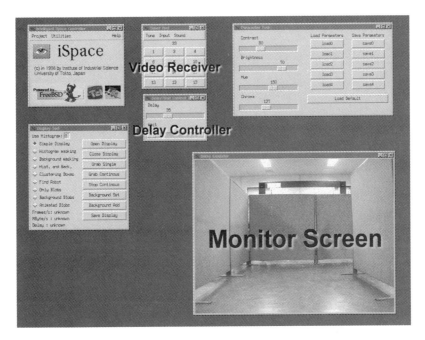

Figure 8.5:   Delay emulator.

8.2.3   Software Architecture

This section describes the software design of the systems. Figure 8.6 shows the mobile robot teleoperation system (system I), which is based on UDP/IP. In the remote site, there are a mobile robot and host PC for a teleconference system. The mobile robot receives velocity, control data, and then PC on the robot controls motors, which drives two wheels. Control data are sent in UDP/IP (User Datagram Protocol on IP) packet continuously.

In the online system (system II), shown in figure 8.7, an image-transfer system was developed. This image-transfer system runs on a web server (Apache on FreeBSD). Captured images by CCD camera are converted to JPEG image, and these images are transferred by NPH (No Parse Headers) script running on the web server with server push method.

To emulate time delay in teleoperation through the Internet, a delay emulation system (system III) and simulator system (system IV) were developed. In the teleoperation delay emulation system (system III), processes until capturing are the

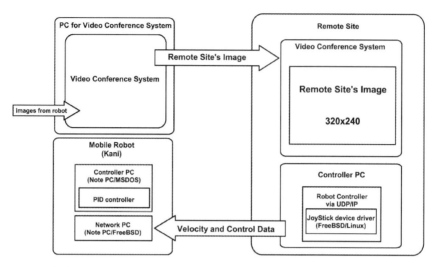

Figure 8.6:   Architecture of system I.

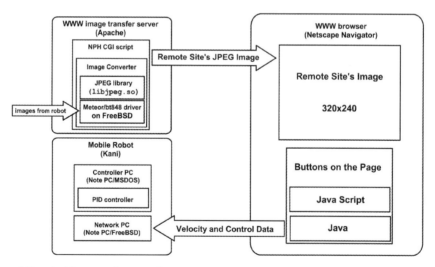

Figure 8.7:   Architecture of system II.

same as system II. Captured images are saved into FIFO and then shown later. This program is written by C/C++ and Tcl/Tk and it runs under FreeBSD. This system can control delay time from 0 to 7seconds and it can also control frame rate. A simple evaluation system was developed to evaluate the influence of time delay in human operators performing teleoperation from point of view of psychological time. This system simulates steering action of mobile robots teleoperation. Graphic user interface is implemented by Tcl/Tk, which is a powerful script language and a GUI tool kit. This system displays only the goal and current position of the robot by red and blue circles, and the operator moves the red circle to the blue circle. This system can also control delay time from 0 to 14 seconds.

8.2.4   Discussion

It is possible that in teleoperations with communication time delay, the instability of operation is caused by human time perception, which is different from the other senses such as visual sensation, auditory sensation, tactile sensation, the sense of taste, and the sense of smell. Moreover, the receptor, which perceives time directly, doesn't exist in the human body. The psychological basis of the time preception of humans has been studied for many years. However, the mechanism for how humans perceive short time is not clear. The mechanism for a human's long time sense, like *circadian rhythms,* is better understood [8].

Until now some understanding about the short time sense has been expressed in *psychological time.*

The Temporal Threshold

Temporal threshold is the shortest time that can be perceived by the sensation [9]. Table 8.1 shows some threshold of time perception in visual, auditory, and tactile sensation. Under the limit, the sensation cannot discriminate between the signals.

Table 8.1: Time perception limit of human

Sensation	Limit
Visual sensation (same one point)	40 msec
Visual sensation (near two points)	Few msec
Auditory sensation	2 msec
Tactile sensation (finger tip)	30 msec

The Psychological Present

The *psychological present* is defined as the time range, which is perceived as present when phenomena happen continuously. It is said that the psychological

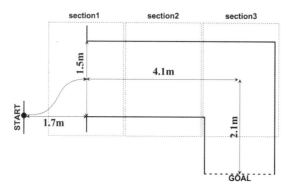

Figure 8.8:   Test track.

present is about 5 to 6 seconds [10] or 2 to 3 seconds [11], which changes according to the mental condition.

Time Estimation

It is said that time longer than the psychological present can not be perceived directly. A human captures the period internally by evaluating or estimating the phenomena that happens [12].

8.3   Performance

This section shows the experimental system setup and experimental results on the performance of human operators, who operate mobile robot teleoperation systems with time delay in systems III and IV.

8.3.1   System Setup

Various time-delay and frame rates of video image were experimented within system III. The experiments were done on the test track, shown in figure 8.8.

The test track consisted of three sections. Section one tested how a mobile robot passes through some object. Obstacle avoidance tasks, which are one of the most basic functions of mobile robots, are equivalent to this task. Section two is a region that evaluates a mobile robot in a parallel motion with some object. In section three the robot moves to a place out of a visual field. This task is equivalent to a task of a change of direction. Width of the passage of the test track is set as two

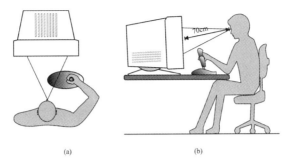

(a)   (b)

Figure 8.9:   Arrangement of the system.

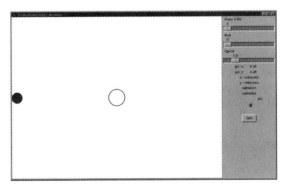

Figure 8.10: Operation display.

times the radius of the circle, which is needed for spinning the robot, and some surplus space. On the other hand, a boundary line between a wall and a floor can be included by an image of camera with this width. The length of section two is long enough for a mobile robot to return to desired path even if the robot leaves it. In this experiment, one of the most important things is that a human operator in the feedback loop could be a stable constant. In the experiments, to eliminate unstable elements such as variance of operation method of an operator or an operator's skill advancement by learning, an operator was trained repeatedly to become an expert saturated with skill. With this method, the essential influence of time delay upon controllability was evaluated.

Next, the experimental system setup of system IV is shown. Figure 8.9 indicates the arrangement of a simulation system, an operator, and a joystick type controller. Figure 8.10 shows the operation display that simulates mobile robot teleoperation. A circle on the left side is a controlled object and the center circle is the goal for the object. This system can emulate time delay from 0.0 to 14.0 seconds in 30msec step. An operator is instructed to move the object to the goal as fast as possible and keep the object on the goal. After an operator keeps an object on a goal for 3 seconds, this goal moves to another place. The maximum speed of the object is about 143 pixels/sec (77 mm/sec) on 19-inch display. An operator chases after the goal again and repeats this operation five times in one session. Furthermore, this session is repeated five times for statistical accuracy. Four subjects who were male and twenty years of age were tested.

In this experiment, an object movement is restricted to one dimension (horizontal). The experimental qualification must be simple to investigate an operator's property in teleoperation with time delay. Subjects are told to practice enough in advance until they learn about operation and its effect on the system. The notable points of the experiment are the following:

- To acquire perceptual-motor coordination (visual-motor coordination), operators must practice the operation sufficiently.

- For stability of experimental data, subjects perform the series of operation five times.

- Data including serious operation errors are removed.

8.3.2   Experimental Results

In the first experiment, position errors from desired path and task completion time with changes of time delay, frame rate of vision image, and velocity of the mobile robot were observed. Figure 8.11 shows relationships between the time delay and the position errors. It can be known that the errors are proportional to delayed time of vision image.

Relationships between the time delay and the task completion time are shown in figure 8.12 with changing velocity of the robot. For up to 1 second of delayed time, the task completion time was not influenced by the time delay. When delayed time exceeds 1 second, however, the task completion time increased suddenly. When the velocity of the robot is 10 cm/second, there were few changes in the task completion time. When the velocity of robot is 20 cm/second or 30 cm/second, however, the task completion time increases proportionally to the time delay. An

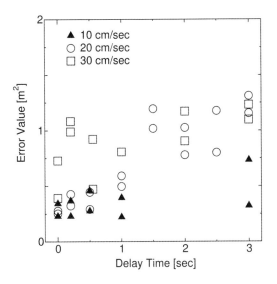

Figure 8.11: Time delay - position error.

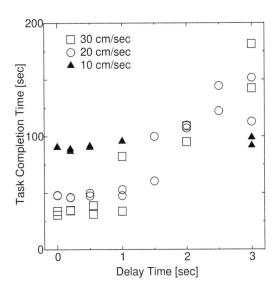

Figure 8.12: Time delay - task completion time.

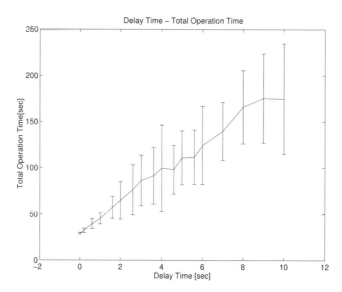

Figure 8.13:   Time delay - task completion time.

extreme change of task completion time is shown when the time delay became larger than 1 second.

When the time delay is about 2 seconds, the task completion time is almost the same on every case of the robot velocity: 10 cm/sec, 20 cm/sec, or 30 cm/sec.

According to these experimental results, guidelines for protocols can be defined to connect local site and remote site in case of teleoperation.

- Choose the best communication medium from available resources.

- Measure communication propagation delay and bandwidth.

- Calculate communication propagation delay for sending image data and control signals.

- According to the delay, set speed of robot and raise robot's intelligent module (if it is available).

- Devices and robots based on these protocols can make connections, which adapt network traffic, bandwidth, and so forth.

Next, the second experiment in system IV will be discussed. Figure 8.13 shows the relationship between delay time and task completion time. The perpendicular

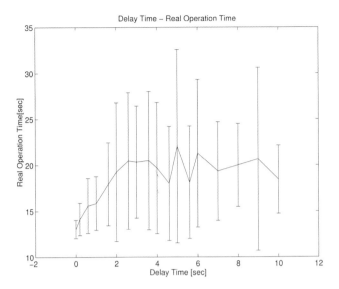

Figure 8.14: Time delay - real operation time.

line in the figure means the deviation of the value. This figure shows that task completion time increases linearly with delay time. Standard deviation of task completion time also increases with delay time. These results indicate that the reproducibility of the operation declines when delay increases. This result includes waiting time, however, since delay time is high, the operator operates in move and wait method.

The relationship between delay time and real operation time is shown in figure 8.14. From this relationship, one can recognize that real operation time increases from 0.0 to 2.0 seconds of delay time. Over 2.0 seconds of delay time, real operation time is almost at same level. Deviation also shows that there is sharp increase from 0.0 to 2.0 seconds and then it gradually decreases from over 6.0 seconds. From these results, it can be shown that the operator changes strategy of operation around 2.0 seconds. This time is related strongly to the psychological present mentioned before. When the time delay is less than 2.0 seconds, the operator can treat it as the present. If it becomes larger, however, the operator cannot treat it as present any longer and should change operation strategy.

Figure 8.15 shows the relationship between delay time and the number of operation times. The number of operation times is defined as how many times the

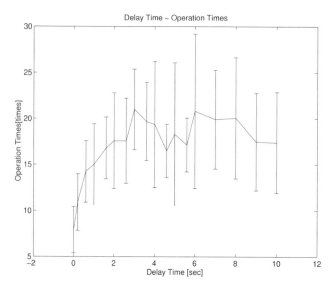

Figure 8.15:   Time delay - number of operation times.

operator inputs control commands through the joystick type controller during the completing the task. This figure indicates that the number of operation times sharply increases from 0.0 to 1.0 second and then it becomes steady. Deviation of the number of operation times increases from 0.0 to 4.0 seconds, and at more than 4.0 seconds, it shows sharp fluctuations. These results also show the relationship between the time delay and the reproducibility. When the time delay became high, the reproducibility was hardly found.

8.4   Conclusions and Future Work

This chapter evaluated the influence of visual feedback time delay and proposed the protocols for mobile robot teleoperation based on experimental results. In addition, the influence of visual feedback in teleoperation was quantitatively evaluated. Three parameters were the focus: time delay of vision image feedback, frame rate of display, and velocity of a robot. We could know what should be prior to others when capacity of communication is limited from the experimental results. Particularly, the relationship among the time delay of vision image, the task completion time, and the velocity of the robot was remarkable (figure 8.12).

According to these experimental results, the protocols to connect local sites and remote sites in case of teleoperation were defined. In future work, protocols for real systems to verify serviceability will be implemented.

The relationship between human behavior and time delay based on psychological time theory in teleoperation system were also explored. Influence of delay time in teleoperation with visual feedback was quantitatively evaluated and reviewed. Experimental results showed some tendencies. When the delay time was from 0.6 to 1.0 second, there is a conversion of operator's operation method to adapt for communication time delay. No apparent results could be found, however, which were significant to design more suitable network connection methods and more usable human interfaces for teleoperation systems. To explain an operator's behavior in teleoperation with time delay, one must analyze the relationship between time delay and human factors. Since the human factor is a significant element in teleoperation, it is worthwhile investigating this subject more deeply.

According to these experimental results, the human interface design methods for teleoperation will be defined. Future work will explore the optimal design of human interface for network robot systems.

Acknowledgments

This project was supported by Kaken of Ministry of Education in Japan. We would like to thank all of the people who have contributed to the project.

Reference

[1]   Y. Kunii and H. Hashimoto. Physical agent and media on computer network. *Proceedings of the fourth International Symposium on Artificial Life and Robotics (AROB)*, vol.1, pp.40 – 45, 1999.

[2]   Noriaki Ando, Joo-Ho Lee, and Hideki Hashimoto. Quantitative Evaluation of Influence of Visual Feedback in Teleoperation. *IEEE International Workshop on Robot and Human Communication*. Takamatsu, Japan, 1998 September.

[3]   Noriaki Ando, Joo-Ho Lee, and Hideki Hashimoto. A Study on Influence of Time Delay in Teleoperation – Quantitative evaluation on time perception and operability of human operator, 1999 IEEE system. *Man and Cybernetics Conference,* Tokyo International Forum, Tokyo, Japan, 1999.

[4]   R.J. Anderson, and M.W. Spong, Bilateral Control of Teleoperators with Time Delay. *Proc. IEEE International Conference System, Man, and*

Cybernetics, vol.1, pp. 131 – 138, 1998.

[5]   Kevin Brady and Tzyh-Jong Tarn. Internet-Based Remote Teleoperation. *IEEE International Conference on Robotics and Automation,* pp. 65 – 70, 1998.

[6]   Ren C. Luo and Tse Min Chen. Supervisory Control of Multisensor Integrated Intelligent Autonomous Mobile Robot via Internet. *IEEE International Conference on Intelligent Processing Systems,* pp. 12 – 16, 1998.

[7]   Kuniaki Kawabata, Tatsuya Ishikawa, Teruo Fujii, Takashi Noguchi, Hajime Asama and Isao Endo. Teleoperation of Autonomous Mobile Robot under Limited Feedback Information. *International Conference on Field and Service Robotics,* 1997

[8]   R.Y. Moore and V. B Eichler. Loss of circadian adrenal corticosterone rhythm following suprachiasmatic lesions in the rat. *Brain Research* 42, pp. 201 – 203, 1972

[9]   L.G. Allan. Temporal order psychometric functions based on confidence-data. *Perception and Psychophysics 18* pp. 369 – 372, 1975

[10] P. Fraisse. Psychological de Temps., 1957, Translated into Japanese by Yoshio Hara, Tokyo Sougen Shinsya, 1960

[11] M. Frankenhaeuser. *Estimation of Time.* Almqvist & Wiksell, 1959.

[12] Fumiko Matsuda, Choushi Kouji, et al. *Psychological Time* (Japanese), Kitaoji Shyobou, pp. 90 – 93, 1996.

9 Personal Tele-Embodiment

Eric Paulos and John Canny

9.1 Introduction

9.1.1 Motivation

Over the past three decades the computer has been transformed from computer-as-exotic-calculator to "digital office," supporting messaging, information organization, and authoring. Electronic mail appeared on ARPANET within a year of its creation, and it soon came to be the dominant form of traffic–networking's first "killer app." Society has watched the growth of other computer-mediated communication (CMC) channels such as USENET news, chat, and instant messaging. More recently, application of Moore's Law[1] has made possible inexpensive desktop videoconferencing and telephony-over-IP. Clearly the computer-as-medium for social communication continues to dominate the net.

Not unlike the introduction of the telephone or telegraph [1, 2], the integration of the computer as a communication tool has already profoundly altered the means of human communication and interaction. Although computers have empowered individuals with new methods of establishing and maintaining contact, it is clear that numerous essential components of human interaction have been lost when compared to "face-to-face" encounters. Many of these components were intentionally discarded, creating unique communication tools adapted to particular tasks. The task-specific choice of media and its corresponding richness is the focus of further discussion by Daft and Lengel [3].

But what exactly has been relinquished during this rapid technological adoption process? Current CMC tools fail to express vital human traits such as instilling trust or employing persuasion [4]. Is it even possible to design systems that encapsulate the richness, warmth, and subtleties of face-to-face encounters? What would such a system look like? How would humans interact across this new

---

[1]. *The observation was made in 1965 by Gordon Moore, cofounder of Intel, that the number of transistors per square inch on integrated circuits had doubled every year since the integrated circuit was invented. Moore predicted that this trend would continue for the foreseeable future. In subsequent years, the pace slowed down a bit, but data density has doubled approximately every eighteen months. This is the current definition of Moore's Law, which Moore himself has blessed.*

medium? How would issues of limited bandwidth and unpredictable latency in current networks be addressed? What will be the new metaphors? Most importantly, is such a system even needed and if so where will it fit into the existing social structure? These questions motivated the author's scientific exploration in the development of robotic and mechanical systems in an attempt to bridge this communication gap. Drawing from current research in computer science, robotics, and social psychology, this chapter traverses several design implementations of such systems.

9.1.2   Related Work

Methods of achieving telepresence[2] are not new with one of the first electrically controlled mechanical teleoperational systems being developed by Goertz [6] in 1954. Since then a variety of applications for teleoperated robotics have been explored [7]. Most of these systems however, are designed for a single specific task and are quite complex. Also, unlike typical telepresence systems employed in remote inspection or hazardous exploration tasks, the primary application here is to facilitate human communication and interaction, called personal tele-embodiment.

The exponential growth of the Web over the past several years has allowed for the development of a plethora of remote-controlled mechanical devices that can be accessed via the Web. The early nonrobotic systems employed fixed cameras in remote spaces where users could observe dynamic behavior such as the consumption and brewing of coffee in a coffee pot [8] or the activity of a favorite pet in its native habitat. Systems evolved to allow users various levels of control via the Web such as the LabCam [9] developed by Richard Wallace. His system allows remote users to aim a pan-tilt camera using an intuitive image-map interface.

Progression to intricate control of more degrees of freedom was realized by introducing robots to the Web. In 1995 Goldberg [10] developed a 3-DOF (degree of freedom) telerobotic system where users were able to explore a remote world with buried objects and, more interestingly, alter it by blowing bursts of

---

2.   *"To convey the idea of these remote-control tools, scientists often use the words teleoperators or telefactors. I prefer to call them telepresences, a name suggested by my futurist friend Pat Gunkel," as quoted by Marvin Minsky in 1980 when discussing the early usage of the term. [5]*

compressed air into its sand-filled world. Another early system was Ken Taylor's "Web Telerobot" [11] which allowed users to stack remote blocks, building elaborate structures over time. Shortly afterward, the authors developed Mechanical Gaze [12], a telerobotic system where web users could control a camera's viewpoint and image resolution to observe various museum artifacts placed within the robot's workspace. Within a short time mobile robots such as Xavier at CMU [13] and Roland Siegwart's Alice robots [14] were connected to the Web. This book serves as an impressive source of material on many of these foundational systems, and thus the reader is referred to the other chapters for more related works in this field.

Although the internet and robotic elements are important in the design of this chapter's systems, it is the social and psychological aspects of CMC that dominate our research. Shared media spaces work such as the PARC Video Whiteboard [15], the ClearBoard [16], and the Media Spaces project [17] have been drawn on. Physical presence and levels of direct manipulation and interaction have also served as inspiration, including Hydra [18], GestureCam [19], Data Dentata [20], and inTouch [21].

9.1.3   Contributions

This work focuses on allowing humans to project their presence into a real remote space rather than a virtual space, using a robot instead of an avatar. This approach is sometimes referred to as "strong telepresence" or "tele-embodiment" since there is a mobile physical proxy for the human at the end of the connection. As a result, the term *tele-embodiment* was coined to emphasize the importance of the physical mobile manifestation [22].

This approach differs fundamentally from more traditional versions of strong telepresence which involve an anthropomorphic proxy or android. Instead, the systems described here attempt to achieve certain fundamental human skills without a human-like form. More important, the research is driven by the study and understanding of the social and psychological aspects of extended human-human interactions rather than the rush to implement current technological advances. To be clear, this is not an attempt to recreate exact face-to-face remote human experiences but rather to extend significantly the richness and abilities of current CMC tools.

Figure 9.1:   (left) A space-browsing blimp in flight, (center) one of the smaller Mylar blimps, and (right) the eye-ball blimp at Ars Electronica 1997.

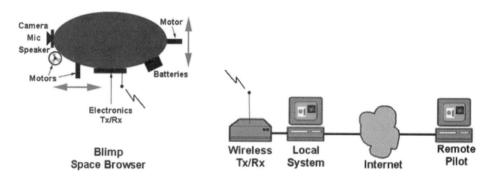

Figure 9.2:   System overview of the basic space-browser (or blimp) hardware configuration.

9.2   Airborne PRoPs: Space Browsing Blimps

9.2.1   Blimp Hardware

The early Personal Roving Presences (PRoPs) were simple airborne telerobots named space browsers first designed and flown in 1995 [22]. A space browser is a helium-filled blimp of human proportions or smaller with several lightweight motors directly connected to small propellers and no other moving parts. On board the blimp are a color video camera, microphone, speaker, simple electronics, and various radio links. The entire payload is less than 600 grams (typically 400 to 500 grams). The design choice here was to use the smallest sized blimps that could carry the necessary cargo, thus making them easily maneuverable down narrow hallways, up stairwells, into elevators, and through doorways. There have been several iterations through different configurations. Blimps ranging in size from 180 x 90 cm to 120 x 60 cm and shapes such as cylinders, spheres, and "pillow-shaped" have all been flown. The smaller blimps consume about the same space as a standing person and are thus well suited for moving into groups of people and engaging in conversation with minimal disruption. Even under full power a blimp moves at human walking pace.

9.2.2   Blimp Software

A user, anywhere on the Internet, uses a Java applet running within a Java-enabled browser to pilot the blimp. As they user guides the blimp up and down or right and left, the blimp delivers, via wireless communications, live video and audio to the pilot's machine. The pilot observes the real world from the vantage of the blimp while listening to the sounds and conversations within close proximity to the blimp. The pilot converses with groups and individuals by simply speaking into the microphone connected to a desktop or laptop computer, the sound delivered via the internet and then a wireless link to the blimp's on-board speaker.

9.2.3   Blimp Discussion

Space browsers are far from perfect in that they require high maintenance. Stringent weight limitations allow for only a small number of batteries to be carried on-board, yielding flights of about an hour before batteries need replacement. Although replacement is quick and rather straightforward, this process still prevents the blimp from operating continuously, as desired. As a result, remote conversations and explorations are often cut short. Furthermore, piloting the blimp is often awkward. Typically the blimp exhibits erratic behavior and the sensation is more like swimming or floating than walking. Another problem is that unlike other robots, blimps are nearly impossible to bring to a complete halt, causing lengthy conversation to be extremely awkward as one participant gracefully floats away in midsentence.

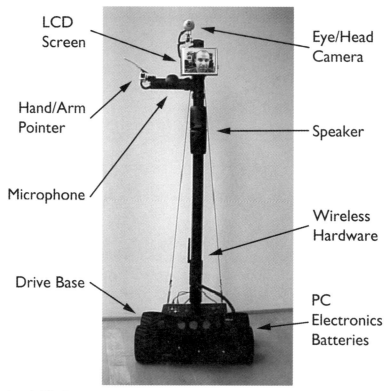

LCD
Screen

Eye/Head
Camera

Hand/Arm
Pointer

Speaker

Microphone

Wireless
Hardware

Drive Base

PC
Electronics
Batteries

Figure 9.3:   A PRoP with camera head, video, LCD screen, controllable arm-hand pointer, microphone, speakers, and a drivable base.

9.3   Terrestrial PRoPs: Surface Cruising Carts

9.3.1   PRoP Hardware

Leveraging off of the previous research with airborne blimp PRoPs, terrestrial four-wheeled surface cruisers or cart PRoPs were developed [23]. These PRoPs were designed from mobile robots with a 1.5 meter vertical pole attached to provide a realistic human vantage for the camera. On board the PRoP are a color video camera, microphone, speaker, color LCD screen, a few simple custom electronics, and various drive and servo motors. Unlike the blimps, PRoPs can travel outdoors, require less maintenance, and provide longer operational time (typically 2 to 6 hours depending on base type and usage). They carry an embedded computer on-board with wireless networking hardware. Booting the PRoP initializes all of its software, and it appears as a server on the network within a couple of minutes.

Dozens of distinct hardware changes have occurred to these ground-based PRoPs during their evolution. Currently, they are constructed from a commercially available robot base with a PC/104 format single board embedded computer onboard running a Microsoft Windows operating system. The computer hardware is an MMX enabled Pentium processor running at 233MHz supporting full-duplex audio and video capture. The PRoP body is a low-weight ABS plastic pole that supports the attachment of various hardware and cabling. Crowning the pole is a high-quality pan-tilt camera that also provides 10x optical zoom, autoiris, and autofocus, all software controlled. Located directly below this camera is a 15 to 20 cm diagonal flat-panel color screen that displays video of the remote PRoP pilot. Off to one side of the screen is the "arm-hand" hardware where a two degree of freedom (2DOF) pointing-gesturing digit-hand is attached using a pair of servomotors. A speaker and omnidirectional surveillance microphone are also located on the stalk of the body to provide the audio channeling. The commercial bases all incorporate simple sonar sensors for taking range measurements. As untethered devices, PRoPs are equipped with enough onboard battery power to support 2 to 6 hours of operational time, depending upon the robot base type. Numerous custom fabricated circuits and in-house C/C++ and assembly language programming of single chip PIC microcontrollers provide the glue from the computer to the various motors, sensors, and control hardware. Ethernet communication uses a standard, commercially available 2.4 GHz Spread Spectrum Wireless interface card fully supporting IEEE 802.11 and operating at 2 Mb/s.

9.3.2   PRoP Software

The software is custom C/C++ code written to interface directly into the H.323[3] videoconferencing standards. This standard has already been incorporated into a variety of existing hardware and software tools and products. Unlike many other

---

3.   *A standard approved by the International Telecommunication Union (ITU) that defines how audiovisual conferencing data is transmitted across networks that cannot guarantee quality of service (i.e., the Internet). By complying to H.323, multimedia products and applications from multiple vendors can interoperate, thus allowing users to communicate without concern for hardware compatibility.*

internet teleoperated systems using one-way video, the system here requires at least two-way audio and video. Therefore, standard server-push images and broadcast-streaming video techniques were not adequate for PRoP usage.

The H.323 standard was adopted because it consists of a suite of custom protocols for transporting multimedia content over unreliable networks such as the internet. More importantly it is a true international standard, widely accepted by numerous hardware and software vendors, designed to intelligently handle dynamic bandwidth rate adaptation, and it is optimized to maintaining sufficient signal quality for human tele-conferencing. Netmeeting is a free software tool from Microsoft that implements the H.323 standard. Here PRoP software utilizes the H.120 data channel component of the H.323 standard for motor control signaling and sensor data flow.

9.3.3   PRoP Interface and Control

Control of the PRoPs is hampered by a traditional design challenge: mapping a low DOF input device into a high DOF output system. For example, a basic PRoP requires at least seven control signals: velocity, steering, head pan, head tilt, pointer pan, pointer tilt, and camera zoom. But the basic input device is a 2DOF mouse. This is not to say that high DOF systems are impossible to control. Almost everyone knows at least one computer gamer who can dexterously control the most complicated spaceship or action hero (or villain) with only a mouse and a few raps on the keyboard. Basic control of PRoPs originated with simple mouse control with keyboard or mouse buttons used for mode selection. This was considerably easier with the blimps where one would rarely need to modify the height and there were no steerable cameras or pointers onboard.

Ground based-PRoP control quickly incorporated a joystick with buttons used for mode selection. Previously difficult tasks such as moving straight ahead or swinging the head back and forth were dramatically improved. The joystick offered tactile feedback on the position of the control signal input about the range of the device (i.e., where the center position was). With the mouse, a user could only determine this visually by looking at the screen controls. Further development incorporated a force-feedback joystick and use of several other input signals found on more modern "gamers" joysticks. This has allowed for easy mapping of "glancing" and camera zooming that would have required additional mode selection techniques, almost certainly complicating the system.

There are still substantial burdens however associated with interfacing into a remote world. Anecdotal observations reveal extreme control difficulty and tedium in two fairly common navigational tasks. The most common tasks are requests of the form "hey, I want to go over there," "I want to get to the end of the hall," or "look, there's Jack, go over there so I can say hello," or "Move over to that door."

One solution currently being explored requires only a single, simple pointing gesture on an image with the mouse or joystick to direct the PRoP. Crisman developed an elegant deictic interface for mobile robot control where users pointed out navigational targets directly in an image [24]. The approach here does not require the use of any image feature target. Instead, from geometric constrains the intersection of the image point is calculated then selected with the ground plane. The control mode will then run closed loop toward the goal without requiring any input from the user. The user can interrupt the system at any time to steer the PRoP to a new location.

The autonomous robot navigation problem is at the core of traditional robotics research and it is not the goal of this work to solve that problem. Recall that PRoPs are not autonomous however but controlled remotely with some limited autonomy to aid with navigation. Therefore, the solution to this problem relies on direct user intervention to assist the PRoP during these extremely difficult navigational situations. The point-and-move interface requires that the goal be directly visible and have an obstacle free path or a path that can be achieved with a simple obstacle following motion planning algorithm. This control will not require actual features to be selected, instead using geometric constraints based on camera properties and position. Again, the objective is to develop a hardware and software solution that will succeed most of the time in automating an extremely common navigation task.

Besides the design of various navigational interfaces, the interested here is also in the construction of simple remembrance maps. These maps can be produced directly from simple sonar measurements and rough odometry data. The Web Driver application [25], designed to aid in real-time web navigation, demonstrates several basic primitives necessary for map building. By capturing high-quality images and embedding them into the visualization of the map interface, a visual history of a televisit is created. This remembrance map data can easily be recalled from storage even when the PRoP is off-line. The hope is that the resulting maps will mimic the episodic style of encounters and events that are typical of the human interaction routine as well as their cognitive remembrance model [26].

9.3.4   PRoP Discussion

PRoPs are designed to enable personal tele-embodiment through the internet. Their usability and success depend primarily on the degree to which they allow humans to communicate and interact. As designers and engineers it is the identification, integration, and interfacing of various communication channels that coalesce to create a compelling, immersive, and expressive experience. A brief description of the various communication cues follows.

- *Two-way audio*: This channel is primarily used for verbal communication as well as sensing subtle background sounds. It is also useful for determining spatial characteristics such as room size, activities, conversations, noise, mood, and so forth.

- *Two-way video*: The video screen allows for exchange of visual cues, such as facial expressions, while the camera serves as a general purpose tool for viewing a plethora of visual data about a remote space (i.e., who's there? what does it look like? is the person I'm talking to even looking at me?). Like the audio data, the video signal undergoes a lossy compression. For a wide range of tasks, humans appear capable of transparently adjusting to this lower quality signal [27]. The H.323 protocol seamlessly allows users to trade frame-rate for image quality depending on the task (navigation verses inspection-observation).

- *Proxemics*[4]: The mobile robotic base allows users to position themselves with a group or individual as well as browse and explore. Thus it permits first-order proxemics to be expressed [28].

- *Directed gaze*: The movable pan-tilt head delivers a gaze direction to remote individuals. Using this nonverbal cue, the PRoP pilot can "turn and face" someone to see them, address them, or just give to them attention. This cue is also extremely important for back channeling during conversations.

- *Deictic gesturing:* A 2-DOF pointer attached near the shoulder of the PRoP facilitates simple pointing and gesturing. Remote users engage this tool to point out a person, object, or direction. Making simple motion patterns, it can express

---

[4.]   *Proxemics is the study of the nature, degree, and effect of the spatial separation individuals naturally maintain (as in various social and interpersonal situations) and of how this separation relates to environmental and cultural factors.*

interest in a conversation, agreement with a speaker, or gain attention for asking a question in a crowded room. To preserve meaning, nuance, and richness in these gestures, continuous input devices are used.

- *Physical appearance, color, and viewpoint:* Employed as a communication tool in a public setting, subtle physical design choices often hinder the effectiveness of PRoPs. Careful attention to their overall height, shape, and color are important. Trial and error as well as user feedback have provided useful anecdotal evidence for and against many of the elements in the current design. A full understanding of the design space of physical PRoP attributes has not yet been gained however and work in this area is ongoing.

9.4   Conclusions and Future Work

PRoPs are telepresence systems designed specifically for human interaction. They are best viewed as media that allow interaction between their pilot and the people in the robot's environment. The emphasis in PRoP design is therefore on capabilities for interaction. PRoPs require less autonomy and intelligence than most robots because they are in continuous control by their pilot. Tasks such as navigation and map-building are synergistic: the robot takes care of sensing, recording, and organizing data, while the pilot provides most of the intelligence in the system in recognition and planning tasks. An overview of some early prototypes of PRoPs was presented here. Current versions are undergoing evaluation now in the first phase of an iterative design cycle. Extensive user testing will be done with these robots. The authors want to characterize the set of tasks that are well supported by the PRoP's current capabilities. From its limitations, we will proceed on the next design iteration.

Acknowledgments

This project was supported by the National Science Foundation under grant numbers FD93-19412 and IIS-9978138. We would also like to thank several individuals for numerous enlightening personal discussions and feedback on this work: Ken Goldberg, Scott Fisher, Ben Zorn, Paul Dourish, Cliff Nass, Hiroshi Ishii, Jason Hong, Mark Newman, James Landay, and Dan Reznik.

References

[1]   C. S. Fischer. *America calling: a social history of the telephone to 1940.* Berkeley: University of California Press, 1992.

[2]   C. Marvin. *When old technologies were new : thinking about electric communication in the late nineteenth century.* New York: Oxford University Press, 1988.

[3]   R. L. Daft and R. H. Lengel. "Information richness: A new approach to managerial behavior and organizational design." *Research in Organizational Behavior*, vol. 6, 1984.

[4]   E. Rocco. "Trust breaks down in electronic contexts but can be repaired by some initial face-to-face contact." 1998.

[5]   M. Minsky. "Telepresence." In *Omni*, vol. 2, 1980.

[6]   R. Goertz and R. Thompson. "Electronically controlled manipulator." *Nucleonics*, 1954.

[7]   T. B. Sheridan. *Telerobotics, automation, and human supervisory control.* Cambridge: MIT Press, 1992.

[8]   D. Gordon and M. Johnson. "Trojan room coffee machine." 1994.

[9]   R. Wallace. "A notebook active vision system." Presented at UI Workshop, 1996.

[10] K. Goldberg, M. Mascha, S. Gentner, N. Rothenberg, C. Sutter, and J. Wiegley. "*Desktop teleoperation via the World Wide Web.*" 1995.

[11] K. Taylor and J. Trevelyan. "A telerobot on the World Wide Web." Presented at the *Autralian Robot Association*, Melbourne, Ausralia, 1995.

[12] E. Paulos and J. Canny. "Delivering real reality to the World Wide Web via telerobotics." 1996.

[13] R . Simmons, R. Goodwin, K. Z. Haigh, S. Koenig, and J. O'Sullivan. "A layered architecture for office delivery robots." 1997.

[14] R. Siegwart, C. Wannaz, P. Garcia, and R. Blank. "Guiding mobile robots through the Web." Presented at *IEEE/RSJ International Conference on Intelligent Robots and Systems, Workshop*, Victoria, Canada, 1998.

[15] J. C. Tang and S. L. Minneman. "VideoDraw: a video interface for collaborative drawing." In *ACM Transactions on Information Systems*, vol. 9, 1991, pp. 170.

[16] H. Ishii, M. Kobayashi, and J. Grudin. "Integration of interpersonal space and shared workspace: ClearBoard design and experiments." in *ACM Transactions on Information Systems*, vol. 11, 1993, p. 349.

[17] P. Dourish, A. Adler, V. Bellotti, and A. Henderson. "Your place or mine? Learning from long-term use of audio-video communication." *Computer Supported Cooperative Work (CSCW)*, vol. 5, pp. 33–62, 1996.

[18] A. Sellen, B. Buxton, and J. Arnott. "Using Spatial Cues to Improve

Videoconferencing." Presented at *ACM SIGCHI*, 1992.

[19] H. Kuzuoka and G. Ishimoda. "Supporting position expressions for spatial workspace collaboration." *Transactions of the Information Processing Society of Japan*, vol. 36, pp. 1379–86, 1995.

[20] K. Goldberg and R. Wallace. "Data Dentata."Presented at *ACM SIGCHI*, 1992.

[21] S. Brave and A. Dahley. "InTouch: a medium for haptic interpersonal communication." Presented at *ACM SIGCHI*, 1997.

[22] E. Paulos and J. Canny. "Ubiquitous tele-embodiment: applications and implications." *International Journal of Human-Computer Studies*, vol. 46, pp. 861–77, 1997.

[23] E. Paulos and J. Canny. "PRoP: Personal Roving Presence." Presented at *ACM SIGCHI*, 1998.

[24] J. D. Crisman, M. E. Cleary, and J. C. Rojas. "The deictically controlled wheelchair" *Image and Vision Computing*, vol. 16, pp. 235–49, 1998.

[25] S. Grange, T. Fong, and C. Baur. "Effective vehicle teleoperation on the World Wide Web." , *Proceedings of IEEE International Conference of Robotics and Automation (ICRA)*, San Francisco, April 2000.

[26] B. J. Rhodes and T. Starner. "Remembrance Agent: a continuously running automated information retrieval system." Presented at PAAM 96. *Proceedings of the First International Conference on the Practical Application of Intelligent Agents and Multi-Agent Technology*, 1996.

[27] B. Reeves and C. I. Nass. *The media equation : how people treat computers, television, and new media like real people and places.* Cambridge University Press, 1996.

[28] E. T. Hall. *The hidden dimension.* 1st ed. Garden City, N.Y.: Doubleday, 1966.

Part III
Remote Control and Time Delay

10 Handling Latency in Internet-Based Teleoperation

Kevin Brady and Tzyh-Jong Tarn

10.1 Introduction

10.1.1 Motivation

Time delays resulting from communication propagation lag have long been stumbling blocks for remote teleoperation. As early as the 1960s, Ferrell [1, 2] and Sheridan [3] started investigating this problem. They found that these delays were potentially destabilizing and proposed the "move-and-wait" strategy for dealing with them. According to this method, the teleoperator sends a move command whose time duration is no longer than the delay. Once the teleoperator observes the motion to be settled, a new command is sent. Needless to say, this is a less than efficient manner in dealing with remote teleoperation.

Historically there are two main arenas of research in remote teleoperation: underwater robotics and space-based robotics. Both incur round-trip delays on the order of seconds when operated remotely. For an excellent review of space teleoperation see Sheridan [4]. More recently, some researchers are beginning to investigate the exciting prospect of utilizing the Internet as a medium for remote teleoperation.

The Internet is well poised as a third major communications medium for facilitating teleoperation. It is much like the other methods in that there are delays that compromise the ability to have a closed-loop control system without the need for further modifications. One difference however, is in the time varying nature of Internet-based communications. Underwater and space-based robotics tend to have delays that are approximately constant, with variation in the delay at least an order of magnitude smaller than the delay itself. The Internet often has delays whose variance is larger than the size of the nominal delay.

The uncertainty of Internet-based communications makes modeling real-time systems quite difficult. There is little research in modeling such systems, though Fiorini [5] and Oboe [6] have several excellent papers discussing the various problems and approaches encountered when using the Internet as a communication medium for control systems.

Internet-based teleoperation is a new concept in search of an application, though some interesting possibilities have been presented. At a plenary session of the 1997 IEEE Conference on Robotics and Automation in Albuquerque, New Mexico, session chair Ray Harrigan of Sandia National Laboratories proposed such teleoperation to represent the future of cooperative research among labs and groups located in remote locations. Laboratories without the physical resources to test out new methods in automation and manufacturing can work cooperatively with remote sites that do, experimenting with such methods without ever leaving the lab.

Another area under investigation is using the Internet as a medium for performing telerobotic surgery [7]. In such a scenario, specialized surgical skills are not available in geographical areas that require such skills. A general surgeon would be available to perform much of the less skilled work, while a surgical specialist would then use a telerobot to perform the more skilled aspects of the surgery.

10.1.2 Previous Work

There have been myriad designs for constructing telerobotic architectures where delays exist in the observation and control of telemanipulators. In general, these attempts can be grouped into three general approaches: predictive control, bilateral control, and teleprogramming. Many designs tend to blur these distinctions.

Predictive control is used here to refer to a broad range of predictive approaches including traditional predictive control [8], Kalman filtering [9], and internal model control [10, 11]. Its role in remote teleoperation is to generate a model for the remote system to compensate for the delay. This predicted model may be used graphically to provide a pseudo real-time response to teleoperator commands. The Jet Propulsion Laboratory (JPL) [12, 13, 14] has done much work in using predictive displays to help control space-based robotic systems. One manifestation of their predictive paradigm is coined the "phantom robot" [12].

Much research has taken place in controlling telerobots using bilateral control. The intention is to control both the contact force and position (velocity) of the telerobot. As a practical matter, doing both simultaneously is not possible. Instead, bilateral control is modeled as a two-port where force is the input and velocity is the output for the teleoperator. Likewise, velocity is the input and force is the output for the telerobot. For more information see [15, 16, 17, 18, 19, 20, 21, 22, 23].

"Teleprogramming" is used here to encompass a broad range of approaches that perform remote teleoperation in a similar way. This approach enables cooperation between the remote planner and the robot on an abstract level to overcome the limitations of delay and bandwidth. The local controller has broad latitude in interpreting the general plan and implementing it. This supervisory control approach typically results in greater modularity and independence for the remote telemanipulator. For more information see [24, 25].

Much research has been done in controlling systems with delays. Such systems have a broad range of applications, including process control, ship stabilization, and many biological problems such as population models. The difficulty is that traditional control methods typically cannot guarantee stability for such systems except for possibly small gains. One way out of this is to take the process model to be controlled and model it as part of the control. The first example of this approach is the well-known Smith predictor [26]. Utilizing a process model, the time-delay is eliminated from the characteristic equation converting the closed-loop system into a delay-free one.

Olbrot [27, 28] established conditions for the controllability, detectability, and stabilization of time-varying delayed systems. Manitius and Olbrot [29] came up with a feedback law to yield a finite spectrum for the closed loop time-delay system. Their research stems from the idea that time-delay systems are infinite dimensional systems. Assuming perfect system knowledge, they develop feedback that includes an integral term on the feedback itself as well as the expected state feedback. Using standard Laplacian methods, they show that the resulting system can be made finitely dimensional, while mapping the finite poles to the left-hand complex plane using standard pole-placement techniques. They further investigate the case of disturbances or imperfect system knowledge and ask two questions. First, does the system become infinitely dimensional? Second, what happens to the poles of the system? Their answer to the first question is, yes, the system does become infinitely dimensional. Their answer to the second question is more interesting. They find that they still have finite poles in the left-hand plane that are in the neighborhood of the poles assigned using standard techniques. Furthermore, the infinite poles come in far out in the left-hand plane. Their conclusion is that the system remains stable for reasonable disturbances-modeling errors.

Watanabe and Ito [30] came up with an observer for the delayed system with finite spectrum, comparing their results with those of the Smith predictor. Klamka [31] extended these results to the case of distributed delays. Later, Pandolfi [32] showed that the Watanabe-Ito observer was a simplified form of the compensator presented in his paper based on the theory of distributed control processes.

Much work has also been done over the years in the area of stability of systems with state delays, particularly the stability independent of the delay magnitude. Li and de Souza [33] use a Razumikhin [34] approach for achieving robust stability of such systems. Kojima et al. [35] reformulate the problem as a H-infinity problem to achieve robust stabilization results. Clarke et al. [36] use an adaptive predictive controller to achieve such ends. Numerous other results exist that come up with stability criterion for time-delay systems independent and dependent on the size of the delay. Other research about systems with state delays include [37, 38, 39].

10.1.3 Contributions

Due to inaccessibility, remoteness, hazardousness, or cost-effectiveness, a human may not always be present in a work environment. Sometimes those humans who are present are not experts (as in the telerobotic surgery case) and they rely on the expertise of a remote specialist. Teleoperation is a viable alternative for projecting human intelligence to these environments. Unfortunately, due to a combination of distance and choice of communication medium, there are frequently communication propagation delays in observation of the remote environment and in control of the remote environment. Such limitation make working in even a structured environment daunting — and structured is seldom the case.

There are two problems. First, how does the teleoperator learn enough about the remote environment to contribute to the task? This is a result of the artificial means that the human relies on to gain such knowledge and is complicated by the observation delay. The second problem is how to project the teleoperator's intelligence to the remote environment, which is complicated by the control delay. The "human in the loop" telerobotic architecture is ideal for closing the gap of uncertainty through use of human intelligence.

Bandwidth limitations and time-delays often limit gaining knowledge of the remote environment. Information that is available is organized and presented to the teleoperator in an easily understood fashion and ideally in real time. If any a priori information on the work cell is available, it is integrated through use of a model. The overall intended effect is to immerse the teleoperator in the remote environment as if he or she were actually present. This ideal is known as *telepresence*.

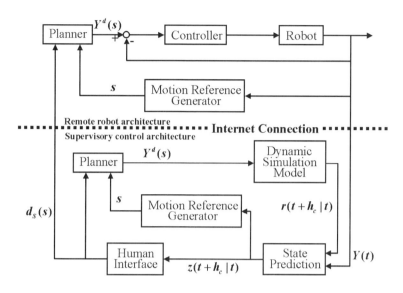

Figure 10.1: Robotic architecture for Internet-based operations with time delays.

The contribution of this work is a general-purpose architecture for projecting "intelligently" into a remote site via a telerobot. It is based on an analytic state-space formulation of the system that includes the appropriate delays in observation and control. An investigation of these delays is a prerequisite to finalizing the state-space model. Upon investigation, these delays often tend to be time varying and nondeterministic. This mathematical analysis is general enough to encompass remote teleoperation using various communication mediums such as sonar, electromagnetic radiation, and the internet.

10.2   Design

10.2.1 Software Architecture and Design

To meet the objectives laid out in the first section there are a number of intermediary goals. These include deriving a flexible supervisory control scheme and developing an interface that gives the teleoperator a sense of telepresence. Figure 10.1 presents the general idea for how to meet the first intermediary goal. It is based on event, based planning and control [40, 41] whose basic design is shown in the top half of figure

10.1. The variable s is the reference variable that is used instead of time, while $Y^d(s)$ and $Y^{ds}(s)$ are the nominal and supervisory commands. The major problem is that the supervisory control is not located where the rest of the system is, and there is a delay in the exchange of information between the supervisory control and the rest of the architecture. The role of this section is to apply the state-space model of the previous section to formulating this problem.

The section of figure 10.1 below the dotted line represents the supervisory controller, while the other half comprise the robot, the controller, and the corresponding planner. The role of the supervisory controller is to deliver a sense of telepresence to the teleoperator. Telepresence is presented through the human interface. The state $z(t - h_c|t)$ that is fed to the human interface represents the fusion of the time-forward observer and a dynamic model of the robot's remote environment. This allows the teleoperator to interact with the virtual environment contained within the human interface as if he or she were doing so in real-time without any delays.

Note that there is an event-based planner and motion reference generator local to each side of the architecture. This distributed architecture gives greater flexibility and autonomy to the remote manipulator. The planned trajectory $Y^d(s)$ is the same on either side. Through use of the time-forward observer presented in the previous section, the reference variable s is coordinated on either side of the system.

TeleSupervisory Commands

Due to communication limitations it is often necessary to communicate with the remote work cell on an abstract level. TeleSupervisory Control, as presented here, aspires to present a canonical methodology for performing such supervisory control. It is unique in a number of ways, including:

- The first robotic teleprogramming technique based on a state-space model.

- Considers the time-varying nature of the communication medium as part of the design.

- It is event based, rather than time based.

This set of supervisory commands is based on the joystick information provided through the human interface. The joystick commanded velocity inputs are discussed in the next section. The supervisory commands are based on function-based sharing control [42] and are extended here for time-delayed systems.

Stop: All motion is suspended:

$$\dot{Y}^d = 0 \tag{10.1}$$

SlowDown: Motion along the nominal trajectory slows down in proportion to joystick input.

$$\dot{Y}^{ds} = \left\{ \begin{array}{ll} \dot{Y}^p_{sp}, & \cos\Theta \leq 0 \\ 0, & \cos\Theta > 0 \end{array} \right\} \tag{10.2}$$

SpeedUp: Motion along the nominal trajectory speeds up in proportion to joystick input.

$$\dot{Y}^{ds} = \left\{ \begin{array}{ll} \dot{Y}^p_{sp}, & \cos\Theta > 0 \\ 0, & \cos\Theta \leq 0 \end{array} \right\} \tag{10.3}$$

Avoid vect: Move off the nominal trajectory by distance *vect*. Note that this may not be possible if the maximum velocity v_m is not sufficient to arrive there in one communication cycle *TC*. The first two steps are performed every *TC* seconds when the new offset arrives, while the last step is performed each time through the control loop while *Avoid* is active. The procedure is as follows:

- Measure the current projected offset of the robot Y^0 from its nominal trajectory.

- Calculate the velocity needed to reach the desired offset from the present offset in one communication cycle *TC*:

$$Y^0_{sp} = \left\{ \begin{array}{ll} \dfrac{vect - Y^0}{TC}, & \left\| \dfrac{vect - Y^0}{TC} \right\| \leq v_m \\[2ex] v_m\left(\dfrac{vect - Y^0}{\|vect - Y^0\|}\right), & \left\| \dfrac{vect - Y^0}{TC} \right\| > v_m \end{array} \right\} \tag{10.4}$$

- At each iteration in the robot control loop the combined command velocity Y^{ds} should be updated as follows:

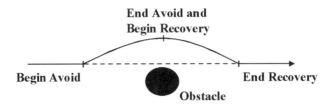

Figure 10.2: Avoid and recover.

Note that there is a corresponding state *(Recover)* for the remote robot to autonomously recover after an *Avoid* command is complete. This whole process is demonstrated in figure 10.2.

$$\dot{Y}^{ds} = \dot{Y}^0_{sp} \tag{10.5}$$

Normal:

$$\dot{Y}^{ds} = 0 \tag{10.6}$$

and several other bookkeeping commands for starting and terminating requests that are further discussed in [43]:

- *Send*: Send the present position of the robot.
- *GoTo position:* Go to position.
- *Start time*: Start robot at time t.
- *Terminate*: Terminate present move.
- *Finished*:
- *Finished_on_recovery*:
- *Finished_on_avoid*:

Time Synchronization

A predictive controller is included in the architecture in figure 10.1. This prediction gives the teleoperator the ability to interact with a virtual model of the remote environment as if there were no time delay. To facilitate this, the observer model for the robot starts its virtual operation h_c seconds before the actual robot starts.

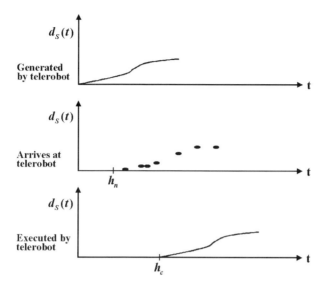

Figure 10.3: Time synchronization of remote and local virtual model.

This time offset gives the teleoperator enough time to react and instruct the remote telerobot to take evasive actions if the teleoperator determines that a collision is about to take place.

Let $t_{observer}$ and t_{robot} represent the local clocks for the observer (local) and robot (remote) work cells, respectively. Each clock starts at 0 when the respective robot (virtual or real) starts execution. The relationship between these clocks at time t is as from the previous paragraph:

The time-forward predictor's goal is for:

$$t_{observer} = t_{robot} + h_c \tag{10.7}$$

If information comes in early, as in figure 10.3, then it is not acted upon until h_c seconds after it was sent.

$$z(t_{observer}|t_{robot}) = x(t_{robot}) \tag{10.8}$$

10.2.2 Interface Design

Figure 10.4 illustrates the human interface that has been successfully tested. It consists of two computer interfaces. The first has a virtual model of the remote

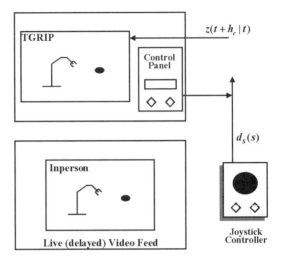

Figure 10.4: Human interface.

system. The human teleoperator may interact with the virtual model in Deneb's TGRIP with the expectation that h_c seconds later the same actions are taking place on the robot. A control panel allows the teleoperator to enable the appropriate functionality of the robot. Live video, using SGI's Inperson that is delayed similarly to the observed data, is shown on a second computer. Finally, a joystick is used to enter supervisory commands.

Internet connection utilizing a UDP/IP link is used for conveying information between the remote sites.

Joystick Input

A joystick velocity input and corresponding command may be provided by the human operator through the human interface. The joystick input corresponds to a desired velocity specified by the human operator. It may be broken down into two parts that are illustrated in figure 10.5 relative to the actual trajectory. One part is the component of the spaceball velocity that is along the present desired trajectory Y^d and another component that is orthogonal to the desired trajectory. Let θ be the angle between Y^d and the spaceball velocity. The components of the spaceball velocity that are parallel and orthogonal to Y^d may be calculated as follows:

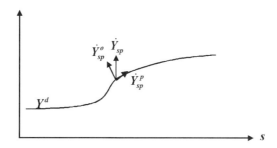

Figure 10.5: Joystick velocity input commands.

$$\dot{Y}_{sp}^{p} = \frac{\dot{Y}}{\|\dot{Y}\|} \left\| \dot{Y}_{sp}^{r} \right\| \cos\theta \qquad (10.9)$$

$$\dot{Y}_{sp}^{o} = \dot{Y}_{sp}^{r} - \dot{Y}_{sp}^{p} \qquad (10.10)$$

10.2.3 Control

Event based planning [44, 45] is used to provide the desired trajectory for this model. It is based on a reparameterization of the control so that it is indexed by an event reference variable s instead of by time. Details are not provided here but are available in the references for those interested in further investigation. These references also provide details of the nonlinear feedback necessary to decouple the robotic system to enable the linear model presented in this subsection.

Time Delays and Bandwidth Limitations

The goal of this section is to describe the nature of the propagation delay that is incurred, describe its relationship to the bandwidth, and come up with a delayed model for the state space system. The unidirectional delay is designated $h(t)$. It is only for illustrative purposes and may denote the delay in either the forward or backward direction. The delay $h(t)$ may be broken down into three components as follows:

$$h(t) = h_{n} + \bar{h}_{d}(t) + h_{b}(t) \qquad (10.11)$$

h_{n} is the nominal propagation delay. It represents the time that it takes the signal to physically propagate without disturbance from its source to its destination

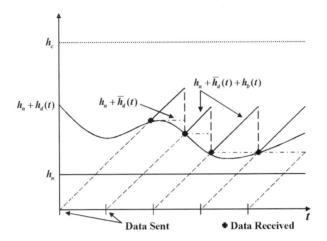

Figure 10.6: Nature of the time delays.

across the communication medium. Its value may be experimentally determined and is non time varying.

h_d is the disturbance delay. It represents the deviation from the expected delay that results from unknown disturbances or even loss of information. h_d is nondeterministic and time varying.

$\bar{h}_d$ is a step function that is based on the function h_d. Since the data exchange is discrete, the only time that the disturbance delay is relevant is when the information actually propagates.

h_b is the bandwidth delay. Since information is exchanged at finite rate b across the communication medium, a corresponding delay will be incurred that is known but time varying. It is not known a priori absolutely, but the form of the function is known, as shown in figure 10.6. The delay h_b is a sawtooth function and is bounded as follows:

$$\bar{h}_b = \frac{1}{b} \geq h_b(t) \geq 0 \qquad (10.12)$$

The choice of b is the first important design consideration of this system. A large value of b enables a larger exchange of information as well as a reduction in the effective delay of the system. On the other hand, there may be bandwidth limitations to just how large a value may be chosen. Choosing b too large may overload the communication system, resulting in lost data and eventually an untimely shutdown.

Note that the delayed information is treated differently depending on if it is sent in the observer or controller direction. Observed information is processed as soon as it comes in. Mathematically, this means that the delay in the observer direction h_o is defined as $h_o(t) = h(t)$. The delay in the control direction is evaluated slightly differently to synchronize the control of the remote robot with a local simulated model. First, a control delay h_c is determined as a design consideration. As mentioned earlier, h_c value represents the difference in time between when a control is applied at the teleoperator and at the telerobot side of the system. It is chosen so that it is at least as large as the maximum value of the deterministic part of the delay model:

$$h_c \geq h_n + \bar{h}_b \tag{10.13}$$

Once the choice of h_c is made a function $\Delta h_c(t)$ can be used to model when $h(t)$ exceeds h_c:

$$\Delta h_c(t) = \left\{ \begin{array}{ll} h(t) - h_c, & h(t) > h_c \\ 0, & h(t) \leq h_c \end{array} \right\} \tag{10.14}$$

$h_c + \Delta h_c(t)$ now represents the time difference between when a control is generated by the teleoperator and enacted by the telerobot. The value of $\Delta h_c(t)$ is bounded by Δh_{max}. If this were not the case, the system would effectively be open loop.

State-Space Model

The state-space model for a telerobotic system with delays in the communication channels is modeled as:

$$\dot{x}(t) = Ax(t) + B_1 u_1(t) + B_2 u_2(t - h_c - \Delta h_c(t)) \tag{10.15}$$

$$y(t) = \sum_{j=1}^{N} C_j x(t - h_{o_j}(t)) \tag{10.16}$$

where $x(t)$ is the task space vector $y(t)$ is the observation of $x(t)$. It is assumed that $u_1(t)$, $u_2(t)$, and $y(t)$ are measurable, though $x(t)$ is not. Matrices A, B_1, B_2, and C are time invariant and of appropriate dimension. The delay components $\Delta h_c(t)$) and $h_{o_j}(t)$ are nondeterministic, time varying, nonnegative, and bounded. The prediction in figure 10.1 is based on this state-space model.

Note that the various components $h_{o_j}(t)$ of the observation delay represent observed information that comes over different channels with possibly different delays. A good example of why such a model is needed is the ROTEX [46] project. NASA allocated two different channels for feeding back the position of the robot and sending back the video feed for the robot. Each of these channels had a different delay. A distributed observer delay model would also be important if there were calculation delays. Rough estimates might be readily available, while more accurate filtered information is available but requires heavy computation and a resulting time lag.

Data Fusion

The architecture's goal is to provide an interface to the robot as if the teleoperator was present in the robot's workspace. A time-forward observer, as presented in the previous section, is an excellent first step to achieving this functionality. The robot however, may physically interact or make contact with the remote environment rather than simply move in free space. A hybrid force-position control is necessary for this reason. To achieve the robust prediction for the hybrid control of the remote manipulator a dynamic simulation of the robot's activities is necessary. This is achieved through fusing the results of the time-forward observer with the results of a dynamic simulation of the remote environment.

The distributed nature of the system's architecture allows the remote telemanipulator to react to unexpected circumstances without needing to wait for a delayed command from the teleoperator. In particular, the event-based planner-controller will suspend forward motion upon contact with an object. This contact cannot be directly modeled in the time-forward observer, though it may be dynamically simulated.

In addition to having a time-forward observer, a dynamic simulation of the remote environment is also an integral design consideration. Let $r(t + h_c | t)$ be the state vector for the simulation representing the state of the delayed system $x(t + h_c | t)$. A simple spring damper system is used to provide a contact model with known objects in the remote environment. The dynamic simulation allows the modeling of contact using these virtual forces. A new state variable

$z(t + h_c|t)$ will represent the weighted average of the time-forward observer and the dynamic simulator. The weights are chosen as follows:

Using these weighting functions one can calculate the estimation for the future state of the robotic system as:

$$0 \leq \alpha(t) \leq 1 \tag{10.17}$$

$$\beta(t) = 1 - \alpha(t) \tag{10.18}$$

$$z(t + h_c|t) = \alpha(t) + r(t + h_c|t) + \beta(t)w(t + h_c|t) \tag{10.19}$$

and the full state prediction scheme is now complete.

In free space the time-forward observer is sufficient for predicting the future state of the remote system. For this case the weightings in the previous subsection may be assigned so that α a goes to 0 and β goes to 1.

Upon collision with an object in the workspace, the state-space model alone is insufficient for completely modeling the interaction. It is necessary to use the dynamic model so that virtual forces may be generated in real time. A virtual force f_v is generated based on the dynamic interaction of the robot and its environment.

In this case it is desirable to have α goes to 1 and β goes to 0. By only commanding the free space motion, and not controlling force, the virtual robot's motion is constrained. If the object is along the robot's trajectory, then the robot will suspend forward motion upon making contact with the object. This is consistent with the ideal of event-based planning. Motion will resume when the object is removed from the path or a supervisory generated to move around the object.

This methodology is similar to Khatib's [47] operational space formulation. He uses an artificial potential field to avoid obstacles and move toward the goal position. In the paradigm presented here, only the former functionality is necessary — creating an artificial force f_v to avoid obstacles.

Note that the parameters α and β serve an important function. In free space the state predictor is sufficient for observing the time-forward state of the robot. The dynamic simulator is not necessary, and the parameters may be set accordingly $(\alpha \sim 1, \beta \sim 0)$. Upon collision the state predictor is incapable of modeling the reflected force, and the dynamic simulator is highly relevant $(\alpha \sim 0, \beta \sim 1)$. It should be noted that if the dynamic simulator was an accurate model of the remote

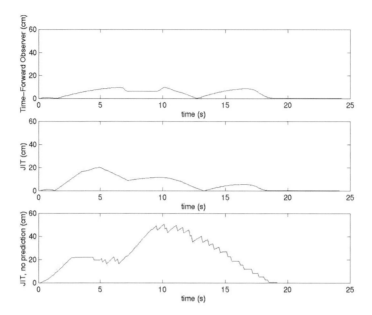

Figure 10.7: Experimental results: move, stop, move.

system, it would be highly relevant at all times and the parameters could be updated to reflect this.

The robot has an a priori model of the remote environment. As the robot model comes near contact with the virtual object, virtual forces are generated. The event-based nature of the controller and planner allows the stable suspension of motion. The TeleSupervisory controller, upon deciding that a collision has occurred issues, a *Stop* command to the remote telerobot.

10.3   Experimental Results

The result shown in figure 10.7 is based on a move along a straight line. The human operator determines that there is an object that may obstruct the move and issues a *Stop* command. After further investigation the maneuver is allowed to proceed. The results are shown based on three approaches:

- Apply supervisory commands h_c seconds after they are issues (approach identified in this research).

- Apply supervisory commands as they show up.

- No prediction is used: supervisory commands are used as they show up.

It is readily apparent that this approach shows a greater correlation between the state of the predictive model and the actual state of the robot. The advantage is that the closer to reality the virtual model is, the more likely that the human operator will attain the intuitiveness (telepresence) for the remote environment to act as an appropriate supervisory controller. Further experimental results are illustrated in reference [43].

10.4   Conclusions

This chapter has presented a promising architecture for controlling remote telerobots. The goals of safety, efficiency, telepresence, and transparency have been met. Its effectiveness has been tested, including a live demonstration during a plenary address at the 1997 IEEE Conference on Robotics and Automation in Albuquerque, New Mexico. Such architecture has ready application to remote-controlled space-based and underwater robots. Additionally, it is highly relevant to the nascent field of Internet-based control.

A delay model was developed and presented. This model is flexible enough to embrace the wide variety of possible communication mediums for remote teleoperation. Limitations in bandwidth and time-varying delays were discussed, and their affect on system design parameters. Based on the delay model, a state-space model for controlling, observing, and predicting the telemanipulator was developed.

Using the state-space model, a corresponding architecture has been developed. A human in the loop architecture was used to ensure the safe and intelligent operation at the other end. This required the development of a supervisory paradigm based on the state-space model. The teleoperator was also immersed in a graphical predictive model to develop a degree of telepresence in the remote environment.

A dynamic model of the robot was fused with the state-space time-forward observer. Using data fusion, the interaction with a virtual object in the remote environment was performed.

Experimental results show the efficacy of this approach; importantly, it is robust to varying time-delays, an area that has not been explored before. The predictive nature of the architecture, as well as the virtual model, allow for a degree of transparency to the user of the remoteness of the working environment and the delays in the communication channels. This is crucial for preventing the degradation of the teleoperator's intuition and performance.

There are a number of future directions for this research. First, work needs to be done in developing more informative graphic and haptic interfaces. The validity of this approach will ultimately be judged by its ability to facilitate tasks in a remote environment. This chapter has presented an analytical framework for modeling the remote environment to compensate for the time-delays. Better user interfaces must be developed to fully take advantage of this approach.

Second, using virtual reality to create a full immersion in the remote environment is highly desirable. In particular, a virtual force-feedback would be effective in immersing the teleoperator in interacting with the remote environment. It would create a greater sense of telepresence. More flexibility in working in the remote environment would also be achieved. This is one of the reasons for integrating a local dynamic simulator into the design. The promising results obtained in predicting the future state of the robot are quite pertinent to force control that is highly dependent on accurate measures of position and velocity.

Third, the framework in this chapter includes the possibility of coordinating multiple arms in working in a remote environment. Experimental work in this area would be beneficial.

The autonomous generation of objects in the system model would be ideal for working in a dynamic environment. Using an intelligent vision and torque system, the dynamic model could obtain a real-time (delayed) model. This could be used to update and model dynamic interactions in the environment.

Sheridan [4] notes the promise of predictive displays as a means for controlling space-based robotics. Intuitively, it is the most practical way to give a sense to the teleoperator of the working situation at the remote sight. He also mentions the promise of shared-compliance control to give an added measure of dealing with uncertainty.

References

[1]   William R. Ferrell. Remote Manipulation with Transmission Delay. *IEEE Transactions on Human Factors in Electronics*, pp. 24–32, September 1965.

[2] William R. Ferrell and Thomas B. Sheridan. Supervisory Control of Remote Manipulation. *IEEE Spectrum*, pp. 91–88, October 1967.

[3] Thomas Sheridan and W. B. Ferrell. Remote Manipulative Control with Transmission Delay. *IEEE Transactions on Human Factors in Electronics*, pp. 25–29, September 1963.

[4] Thomas Sheridan. Space Teleoperation through Time Delay: Review and Prognosis. *IEEE Transactions on Robotics and Automation* 9 (5): 592– 606, October 1993.

[5] Paolo Fiorini and Roberto Oboe. Internet-Based Telerobotics: Problems and Approaches. *International conference on Advanced Robotics*. Monterey, CA, USA.

[6] Roberto Oboe and Paolo Fiorini. Issues on Internet– Based Teleoperation. *IFAC SYROCO*, Nantes, France, September 1997.

[7] Alberto Rovetta, Remo Sala, Xia Wen, and Arianna Togno. Remote Control in Telerobotic Surger. *IEEE Transactions on Systems, Man, and Cybernetics – Part A: Systems and Humans* 26 (4): 438– 444, July 1996.

[8] Ronald Soeterboek. *Predictive Control: A Unified Approach*. Prentice Hall, 1992.

[9] Ian b. Rhodes. A Tutorial Introduction to Estimation and Filteing. *IEEE Transactions on Automatic Control*, AC-16 (6): 688–706, December 1971.

[10] Naoto Abe. Practically Stability and Disturbance Rejection of Internal Model Control for Time-Delay Systems, *Conference on Decision and Control*.

[11] Ning-Shou Xu and Zhi-Hong Yang. A Novel Predictive Structural Control Based on Dominant Internal Model Approach. *IFAC*, San Francisco, pp. 451– 456, 1996.

[12] Antal K. Bejczy, Paolo Fiorini, Won Soo Kim, and Paul S. Schenker. Toward Integrated Operator Interface for Advanced Teleoperation Under Time-Delay. *Intelligent Robots and Systems* pp. 327–348, 1995.

[13] Antal K. Bejczy, Won. S. Kim, and Steven C. Venema. The Phantom Robot: Predictive Displays for Teleoperation with Time Delays. *IEEE International Conference on Robotics and Automation* Cincinatti OH, pp. 546–551, May 1990.

[14] Won S. Kim and Antal K. Bejczy. Demonstration of a High-Fidelity Predictive / Preview Display Technique for Telerobotic Servicing in Space. *IEEE Transactions on Robotics and Automation* 9 (5): 698–701, October 1993.

[15] Robert J. Anderson. Teleoperation withVirtual Force Feedback. *Prodeecings of the 1993 SPIE International Symposium on Optical Tools for Manufacturing and Advanced Automation* Boston MA, September 1993.

[16] Robert J. Anderson. How to build a Modular Robot Control System using Passivity and Scattering Theory. *IEEE Conference on Decision and Control* New Orleans, December 1995.

[17] Gunter Niemeyer and Jean-Jacques E. Slotine. Shaping the Dynamics of Force Reflecting Teleoperation with Time-Delays. *IEEE International Conference on Robotics and Automation, Workshop on Teleoperation and Orbital Robotics.*

[18] Dale A. Lawrence. Stability and Transparency in Bilateral Teleoperation. *IEEE Transactions on Robotics and Automation* 9 (5): 624–637, October 1993.

[19] H. Kazerooni, Tsing-Iuan Tsay, and Karin Hollerbach. A Controller Design Framework for Telerobotic Systems. *IEEE Transactions on Control Systems Technology* 1 (1): 50–62, March 1993.

[20] Joseph Yan and S. E. Salcudean Teleoperation Controller Design Using H-infinity–Optimization with Applications to Motion Scaling. *IEEE Transactions on Control Systems Technology* 4 (3): 244–258, May 1996.

[21] Gary M. H. Leung, Bruce A. Francis, and Jacob Apkarian. Bilateral Controller for Teleoperators with Time Delay via μ-synthesis. *IEEE Transactions on Robotics and Automation* 11 (1): 105–116, February 1995.

[22] Won Kim. Force-Reflection and Shared Compliant Control in Operating Telemanipulators with Time-Delay. *IEEE Transactions on Robotics and Automation* 8 (2): 176–184, April 1992.

[23] Sukhan Lee and Hahk Sung Lee. Modeling, Desing, and Evaluation of Advanced Teleoperator Control Systems with Short Time Delay. *IEEE Transactions on Robotics and Automation* 9 (5): 607–623, October 1993.

[24] Lynn Conway, Richard A. Bolz, and Michael W. Walker. Teleautonomous Systems: Projecting and Coordinating Intelligent Action at a Distance. *IEEE Transactions on Robotics and Automation* 6 (2): 146–158, April 1990.

[25] Y. Wakita, S. Hirai, K. Machida, K. Ogimoto, T. Itoko, P. Backes, and S. Peters. Applications of Intelligent Monitoring for Super Long Distance Teleoperation. *IROS*. Osaka, Japan, November 1996.

[26] O . J. M. Smith. A Controller to Overcome Dead Time. *ISA* 6: 28–33, 1959.

[27] Andrzej W. Olbrot. On Controllability of Linear Systems with Time Delays in Contol. *IEEE Transactions on Automatic Control* pp. 664–666, October 1972.

[28] Andrzej W. Olbrot. Stabilizability, Detectability, and Spectrum Assignment for Linear Autonomous Systems with general Time Delays. *IEEE Transactions on Automatic Control* AC-23 (5): 887–890, October 1978.

[29] Andrzej Z. Manitius and Andrzej W. Olbrot. Finite Spectrum Assignment Problemfor Systems with Delays. *IEEE Transactions on Automatic Control* AC-24 (4): 541–553, August 1979.

[30] Keiji Watanabe and Masami Ito. An Observer for Linear Feedback Control Laws of Multivariable Systems with Multiple Delays in Controls and Outputs. *Systems and Control Letters* 1 (1): 54–59, July 1981.

[31] Jerzy Klamka. Observer for Linear Feedback Control of Systems with

Distributed Delays in Controls and Outputs. *Systems and Control Letters* 1 (5): 326–331, 1982.

[32] Luciano Pandolfi. Dynamic Stabilization of Systems with Input Delays. *Automatica* 27 (6): 1047–1050.

[33] Xi Li and Carlos E. de Souza. Criteria for Robust Stability of Uncertain Linear Systems with Time-Varying State Delay. In *Thirteenth IFAC World Congress*. San Francisco, pp. 1–6, June 1996.

[34] Jack K. Hale and Sjoerd M. Verduyn Lunel. *Introduction to Functional Differential Equations.* Springer-Verlag, 1993.

[35] Akira Kojima, Kenko Uchida, Etsujiro Shimemura, and Shintaro Ishijima. Robust Stabilizatin of a System with Delays in Control. *IEEE Transactions on Automatic Control* 39 (8): 1694–1698, August 1994.

[36] D. W. Clarke, E. Mosca, and R. Scattolini. Robustness of an Adaptive Predictive Controller. *IEEE Transactions on Automatic Control* 39 (5): 1052–1056, May 1994.

[37] S. D. Brierley, J. N. Chiasson, E. B. Lee, and S. H. Zak. On Stability Independent of Delay for Linear Systems. *IEEE Transactions on Automatic Control* AC-27 (1): 252–254, February 1982.

[38] Takehiro Mori. Criteria for Asymptotic Stability of Linear Time-Delay Systems. *IEEE Transactions on Automatic Control* AC-30 (2): 158–161, February 1985.

[39] Takehiro Mori and H. Kokame. Stability of x(t)=ax(t) + bx(t-τ). *IEEE Transactions on Automatic Control* 34 (4) 460–462, April 1989

[40] Ning Xi. Event-Based Motion Planning and Control for Robotic Systems. Ph.D. thesis, Washington University, September 1993.

[41] Ning Xi, Tzyh-Jong Tarn, and Antal K. Bejczy. Intelligent Planning and Control for Multi-Robot Coordination–An Event-Based Approach. *IEEE Transactions on Robotics and Automation* 12 (3): 439–452, June 1996.

[42] Chuanfan Guo, Tzyh-Jong Tarn, Ning Xi, and Antal K. Bejczy. Fusion of Human and Machine Intelligence for Telerobotic Systems. *Proceedings of the IEEE/RSJ International Conference on Robotics and Automation* pp. 3110–3115, August 5–9, 1995.

[43] Kevin Brady. Time-Delayed Telerobotic Control. Doctor of Science *Ph.D. Thesis*, 1997.

[44] Ning Xi. Event-Based Motion Planning and Control for Robotic Systems. *Ph.D. Thesis*, Washington University, September 1993.

[45] Ning Xi. Tzyh-Jong Tarn, and Antal K. Bejczy. Intelligent Planning and Control for Multi-Robot Coordination–An Event-Based Approach. *IEEE Transactions on Robotics and Automation* 12 (3): 439–452, June 1996.

[46] Gerd Hirzinger, K. Landzettel, and Ch. Fagerer. Telerobotics with Large Time-Delays–the ROTEX Experience. *IROS*, Munich, Germany, pp. 571–578, September 12–16, 1994.

[47] Oussama Khatib. Real-time Obstacle Avoidance for Manipulators and Mobile Robots. *International Journal of Robotics Research* 5 (1): 90–98, 1986.

11 Toward Bilateral Internet Teleoperation

Günter Niemeyer and Jean-Jacques E. Slotine

11.1 Introduction

The Internet is revolutionizing the way people communicate and interact, not just because it can efficiently transmit large amounts of data over vast physical distances but because it can do so in an individualized, private, and most importantly interactive fashion. This has created a virtual world where people meet, conduct business, and share ideas and information without the real world's physical limitations. But as efficient and flexible as this virtual world has become, people fundamentally live in physical space and so use scanners and printers, microphones and speakers to bridge the physical senses with this virtual world. People can thus see and hear into the virtual world, or via the Internet to far away places.

Robotics can further bridge the virtual world with the ability to manipulate physical objects. Connecting a robot to the Internet can ultimately allow anyone to interact physically with the real world or other people from anywhere else. For example, many webcams use movable (robotic) platforms to enable the user to reorient the camera and change views. More interestingly, the TeleGarden project used an Adept robot for web users to interact with a remote garden, planting, watering, and tending to living plants (http://telegarden.aec.at/). And, of course, remotely operated robot vehicles can be directed from far away, with the Mars Pathfinder microrover being perhaps the most extreme example [2].

At the same time, haptics can add the sense of touch and make an experience more real to the user. This can be seen in joysticks and electronic games, where force feedback makes the action appear more real. The same is true for training devices, such as flight simulators. Other complex modeling or data presentation applications can benefit from the ability to feel things as well as see them. Various haptic displays and active joysticks are becoming more popular. The Phantom device, for example as shown in figure 11.1, provides force feedback in three translational directions.

Combining these elements — haptics, robotics, and the Internet — provides the tools to build true teleoperation systems, where users anywhere can not only see and hear but also feel and interact with remote objects or one another. Imagine

Figure 11.1: *The Phantom haptical device provides the user with a penlike handle, seen on the left side of the image. Three motorized axes generate force sensations in the three translational directions at the tip of this handle, allowing the mechanism to simulate arbitrary interactions at that point.*

the e-mail message that actually shakes your hand, or the ability to feel the fabric of the dress you are about to order on line.

Teleoperation has enjoyed a rich history and has always inspired visions of interacting with environments far removed from the user [8]. Indeed the notion of telepresence suggests reproducing all senses at a distance so the user feels, hears, sees, and perhaps even smells the remote objects as if the user were actually there. It is the expansion of the Internet that takes these ideas from a small scientific community to the general public.

Using existing non-Internet teleoperators as a guide, the inclusion of force feedback and haptics with the ability to feel remote sensations greatly enriches the experience and improve the user's ability to perform complex tasks [9]. Such a closed-loop or bilateral setup, however, where the requested action directly effects the resulting sensation and the observed sensation will alter the next action, is sensitive to delays in data transmission. If one can not feel an impact until a considerable time later, then the intermediate actions may cause significant damage, or worse, the system may become unstable when actions and reactions are not properly synchronized. Through the introduction of the wave variable concept

[4], based on a reformulation of the passivity formalism of [1], systematic analysis tools can be developed to understand these problems. Moreover, using wave variables for data encoding, stable force reflecting teleoperators can be designed and shaped to act as simple virtual tools when exposed to large delays. They are also transparent to the user when delays are below the human reaction time [5].

Internet teleoperators further compound these stability problems, as the transmission delays are unpredictable and highly variable. This work expands the wave variable formulation to handle such delay fluctuations, incorporating information as soon as it is available but modulating signals if necessary to assure proper behavior. For constant delays this scheme upgrades to the standard approach and hence provides a safety blanket against unknown or unexpected network delays.

While these developments have been derived and tested in the context of current robotics and haptics technologies, which typically interact with the environment or user via a single or a few points of contact, they fundamentally address the issues of bilateral remote connections with data transmissions over an uncertain network. As such they will be all the more important as interface technologies expand to use gloves, whole body interfaces, and other immersive concepts.

Section 11.2 briefly describes these experiments as well as the observed network transmission effects. Section 11.3 then reviews the basic wave variable approach. Section 11.4 describes the problems of variable delays, which lead to the reconstruction filter scheme presented in section 11.5. The chapter concludes with some final remarks in section 11.6.

11.2   Experiments in Internet Teleoperation

To demonstrate the techniques described below and properly understand the effects of Internet-based transmission, the following experimental setup was constructed, depicted schematically in figure 11.2. Use a standard Phantom device (see figure 11.1) as the haptics or master device, which the user will control. Use a second Phantom device as the robotic or slave device, removing the user's handle so that it can interact directly with the environment. Though both devices (master and slave) were located at MIT, the data was transmitted from the master computer, over the Internet via a distant routing computer, to the slave computer and back. By

Local Computer   Remote Computer

Figure 11.2: The Internet teleoperation experiments have the user holding a haptic device attached to the local computer, a robot controlled by the remote computer, and both computers communicating via the Internet. Additional information and video clips of the experiments can be found on the web at http://web.mit.edu/nsl/www/teleoperation/.

selecting different routing computers in various physical locations ranging from across town to Australia, this setup was able to truly execute and evaluate short-distance and long-distance Internet teleoperation.

Using three degree of freedom Phantom devices allows free motion in space as well as contacting and moving various objects. With the data routed only short distances and correspondingly short delays, the system appears transparent with crisp force feedback, delivering a good sense of telepresence. As either the data transmission is routed over longer distance or as network overloads trigger longer delays, the system appears more compliant and objects seemingly receive a rubber coating. Such an automatic performance reduction is as temporary as the network problems and the crisp feeling is restored when the transmissions speed up. In effect, the instantaneous closed-loop bandwidth is inversely proportional to the instantaneous round-trip delay.

Other experiments simulated complete network failure by unplugging the Internet connection, causing the system to go limp. Replugging the connection automatically reestablishes communications and inherently realigns the master and slave system, even while the user is moving. And finally, attempting to transmit data improperly, without the benefit of wave variable encoding, led to gross instability with the mechanisms shaking violently beyond the user's ability to dampen. This clearly underlines the need for the implicit safety net that wave variables provide.

11.2.1 Network Transmission Effects

When information is transmitted over the Internet, it is divided into small packets and routed in real time through a possibly large number of intermediate stops.

While the average latencies may be low, the instantaneous delays may increase suddenly due to rerouting or other network traffic. In the extreme, the connection may be temporarily blocked. Such effects can distort a signal (as seen with respect to time), can introduce high-frequency noise, and can lead to instability if left untreated.

Table 11.1:   *Approximate round-trip delay times based on physical distance. Note that many other factors also effect the observed delay, from the type of network connection, the time of day, to the data size transmitted.*

Physical distance	Round-trip delay times (msec)
Same local network	5
Same building	20 – 50
Same city	~ 50
MIT (Massachusetts) to California	~ 100
MIT (Massachusetts) to Europe	150 – 200
MIT (Massachusetts) to Japan	300 – 350
MIT (Massachusetts) to Australia	~ 400

Table 11.1 shows the average latencies observed when transmitting data over different physical distances. Note these delay times may fluctuate in large fashion due to many other factors, for example, the time of day or the network load. But in general the latencies are comparable to the human reaction time, at least when transmitting within the U.S. This implies that their effect can be made largely transparent as was confirmed in the experiments.

The instantaneous variations in the delay may be quite strong, however, changing rapidly or often discontinuously. Such "noise" or wobble contains many frequency components similar to the signal itself, so that it may interact with the actual system and ultimately cause instability. Figure 11.3 shows six graphs (a through f) of observed delay times, covering different physical distances and showing different effects. Each circle represents the arrival of a packet of data and is plotted at the time it arrives versus the time it was traveling. The effects observed include:

- *Noisy delay* The amount of time a packet requires to travel between two computers on the Internet will vary in unpredictable and often significant

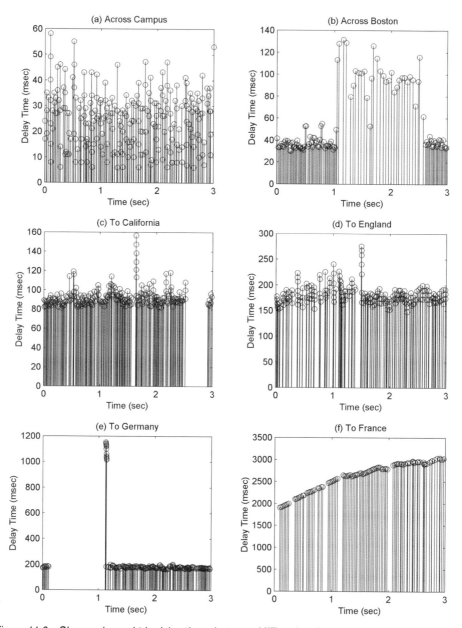

Figure 11.3: *Observed round-trip delay times between MIT and various locations. The graphs not only show the magnitude of the delay but also different effects that the Internet can exhibit.*

fashion. This is true in all cases but most noticeable when the delay itself is small, as seen in graph (a).

- *Variable delay* Based on the time of day and other factors, the network load may significantly change the delay. This may happen during the course of the day, or suddenly, as seen in graph (b).

- *Limited data rates* The amount of data transmitted is inherently limited. More precisely, data are not transmitted continuously but collected into packets and transmitted at discrete intervals. And the rate at which packets are transmitted is limited. Transmitting too much data or too frequently can overload the transmission and cause delays.

- *Bunching* Due to the variable delay at which packets travel, they sometimes arrive in quicker succession than originally transmitted. Furthermore, if packets are transmitted too frequently, the network may group them and deliver them in bigger bunches at a slower rate. Notice this is in the first half of graph (d) and in the middle of graph (c), where multiple circles occur (arrive) at the same time.

- *Temporary blockage and bursting* Due to rerouting, overloading, or other reasons, the network sometimes blocks transmission for considerable time and then delivers data much later than expected, sometimes in huge chunks. This is most evident in graph (e).

- *Out of order, bad, or lost data* Especially if the network reroutes packets due to congestion, they may arrive in a different order than originally transmitted. Also, very infrequently, packets are corrupted or lost completely. This may be detected by numbering packets sequentially and using check sums. These effects are circumvented by ignoring such packets so that they do not appear in the graphs.

- *Other effects* This is clearly not a complete list; for example an extremely slow and slowly increasing delay was observed during a connection to France in graph (f). Fortunately, most other slow effects will have little impact if the system can tolerate arbitrary delays.

11.2.2 Proposed Transmission Scheme

To minimize any effect of Internet transmission on a teleoperator, the following rules are proposed. They are based on the realization that transmission of continuous signals defining a system state (as the teleoperator requires) is inherently different from transmission of files. Most importantly, once a valid set of data has arrived, all previous data are irrelevant and can be ignored. Thus retransmission of lost data is pointless when newer data are already available. In contrast, file transmission requires all data in the correct order.

The rules for signal transmission can be summarized as:

- Combine the current samples of all transmitted signals (thus defining the state of the transmitted variables) into a single packet. This assures that all signals are delayed equally and are self-consistent with one another. It also avoids the need to wait for multiple packets before reconstructing the signals.

- Use small data packets. In particular, transmit auxiliary information and images separately, as large transfers are automatically broken into smaller packets that need to be reconstructed and wait for each other at the receiving site.

- Number the packets sequentially and incorporate check sums. This allows the receiving site to verify order and validity of incoming packets.

- Ignore all but the most current valid packet, meaning corrupted packets, out of order packets, and missing packets. If multiple packets are received simultaneously, use only the latest.

- Use a low-level protocol, in particular UDP, instead of TCP. The higher level protocol will incorporate automatic retransmission of missing or corrupted data, reordering of packets and other mechanisms to guarantee that all packets are received in order. Such sophistication is clearly useful for file transmissions, but leads to higher transmission delays and variability and is not effective for signal transmission.

11.3   Wave Variables

Wave variables are central to these developments and are briefly reviewed here (see [5, 6] for a detailed discussion).

11.3.1   Definition

The key feature of wave variables is their encoding of velocity $\dot{x}$ and force F information, as seen in their definition

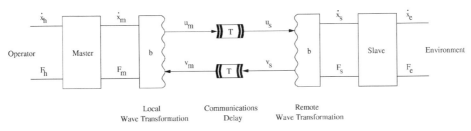

Figure 11.4: The wave-based teleoperator transforms both local and remote information into wave variables before transmission to the other side.

$$u = \frac{b\dot{x} + F}{\sqrt{2b}} \;\; ; \;\; v = \frac{b\dot{x} - F}{\sqrt{2b}} \tag{11.1}$$

Here u denotes the forward or right moving wave, and v denotes the backward of left moving wave. The characteristic wave impedance b is a positive constant or a symmetric positive definite matrix and assumes the role of a tuning parameter, which allows matching a controller to a particular environment or task.

To clarify this definition, consider the typical layout of a teleoperator system shown in figure 11.4. At the local (master) side, the velocity $\dot{x}_m$ is observed and together with the left moving (feedback) wave v_m provides both the right moving (command) wave u_m and the force feedback signal F_m.

$$u_m = \sqrt{2b} \cdot \dot{x}_m - v_m \;\; ; \;\; F_m = b\dot{x}_m - \sqrt{2b} \cdot v_m \tag{11.2}$$

The command wave signal u_m is transmitted to the remote (slave) side where it arrives after the delay T.

$$u_s(t) = u_m(t - T) \tag{11.3}$$

There, together with the observed slave force F_s, it is decoded into a velocity command $\dot{x}_s$ and the feedback wave signal v_s.

$$v_s = u_s - \sqrt{\frac{2}{b}} \cdot F_s \;\; ; \;\; \dot{x}_s = \sqrt{\frac{2}{b}} \cdot u_s - \frac{F_s}{b} \tag{11.4}$$

Finally the feedback wave v_s is transmitted back to the master and arrives after another delay period.

$$v_m(t) = v_s(t - T) \tag{11.5}$$

Note that this layout assumes a velocity command at the slave site and a force-feedback command at the local master site. The wave variable definition however, is symmetric in that either site may decode the incoming wave as a velocity or force command as needed.

This symmetry between force and velocity and the inherent combination of these signals makes a wave system well suited for interaction with unknown environments. Indeed, it behaves like a force controller when in contact with a rigid object and like a motion controller when in free space. The parameter b also allows online trade-off between the two quantities, thus fine tuning the behavior.

A wave signal itself is best described as a command to move or push command, where the sign determines the direction. The receiving side must either move or apply forces depending on its current situation. If no motion was possible, the returning wave will have the opposite sign, signalling that it is pushing back. Or if motion occurred, it will have the same sign indicating that it is moving with the command.

11.3.2 Passivity

The motivation and major benefit of encoding information in wave variable format for transmission is the effect this has on passivity. Indeed, the power flow at any location may be computed as

$$P_{flow} = \dot{x}^T F = \frac{1}{2}u^T u - \frac{1}{2}v^T v \tag{11.6}$$

where power flowing to the right is considered positive. Notice that each wave signal effectively contains its own power needed to execute its action, independent of its opposing or dual wave. Indeed, the power of an individual wave is given by half of its value squared. This is in contrast to the power variables $(\dot{x}, F)$ which must be multiplied together, making it impossible to determine the power required for or contained within an individual signal. The independence between dual wave signals is what makes them robust to delays and so well suited for transmission and teleoperation.

Remember the condition for passivity, which requires that more energy enter a system than be returned from it

$$\int_0^t P_{in} d\tau \geq -E_{store}(0) \qquad\qquad \forall t \geq 0 \qquad\qquad (11.7)$$

where $E_{store}(0)$ denotes the initial stored energy [10]. For wave variables this can be rewritten as

$$\int_0^t \frac{1}{2} v^T v d\tau \leq \int_0^t \frac{1}{2} u^T u d\tau + E_{store}(0) \qquad \forall t \geq 0 \qquad\qquad (11.8)$$

if one assumes that u describes the input wave and v the output wave. Not surprisingly, the output wave energy must be limited to the input wave energy plus any initially stored energy.

Now, if the output wave is simply a delayed duplicate of the input wave, then such an element is passive and furthermore simply stores the wave energy for the duration of the delay. Consider again the basic teleoperator layout of figure 11.4. Assume for now a constant delay in the communications between local and remote sites. The total power input into the communications block at any point in time is given by

$$P_{in} = \dot{x}_m^T F_m - \dot{x}_s^T F_s \qquad\qquad (11.9)$$

with positive master power flowing into and positive slave power flowing out of the block. Using wave variables, one can also compute this as

$$P_{in} = \frac{1}{2} u_m^T u_m - \frac{1}{2} v_m^T v_m - \frac{1}{2} u_s^T u_s + \frac{1}{2} v_s^T v_s \qquad\qquad (11.10)$$

where all variables are measured at the current time t. Substituting the transmission laws from equations (11.3) and (11.5), we find that the input power is stored according to

$$E_{store}(t) = \int_0^t P_{in} d\tau = \int_{t-T}^t \left(\frac{1}{2} u_m^T u_m + \frac{1}{2} v_s^T v_s \right) d\tau \geq 0 \qquad\qquad (11.11)$$

assuming zero initial conditions. As expected, the wave energy is temporarily stored while the waves are in transit, making the communications not only passive but also lossless. This is independent of the delay time T and furthermore does not require knowledge thereof.

Readers are referred again to [5] or their own derivation to show that such a passivity and robustness argument can not be made when transmitting power variables directly, as is the case in most classical teleoperator designs. This is the reason that most classical designs are sensitive to transmission delays and become unstable with unlimited energy being generated inside the transmission elements.

Finally, it has recently been shown that a wave-based scheme is not only passive, but actually contracting [3]. This implies that it exponentially forgets initial conditions or temporary disturbances, with a time-constant equal to that of the slowest of the master or slave subsystem.

11.3.3 Position Tracking with Wave Integrals

In the above basic form, a wave-based teleoperator transmits the wave signals, which encode both velocity and force but do not contain any explicit position information. Position information is thus never exchanged between the two sites and position tracking is only achieved if velocity commands are accurately integrated into position commands and tracked accordingly.

To provide explicit position feedback, one expands the approach to use and transmit the wave integrals in addition to and in parallel with the wave signals themselves. Indeed, just as the wave signals encode velocity and force, their integrals encode position and momentum information.

The integrated wave variables are defined as

$$U(t) = \int_0^t u(\tau)d\tau = \frac{bx + p}{\sqrt{2b}} \quad ; \quad V(t) = \int_0^t v(\tau)d\tau = \frac{bx - p}{\sqrt{2b}} \qquad (11.12)$$

where x denotes position and p denotes momentum, which is the integral of force

$$p \int_0^t F d\tau \qquad (11.13)$$

Note in most cases, little or no importance is placed on the actual momentum value and it can often be eliminated from the system if so desired. Also transmitting the wave integrals together with the wave signals does not change the passivity

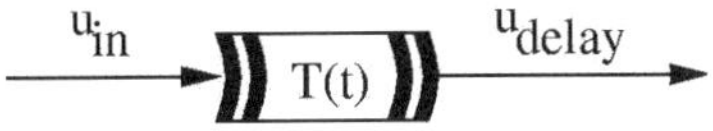

Figure 11.5: Variable time delay in the wave domain.

arguments. It only provides an explicit feedback path and hence robustness for what is already achieved implicitly.

11.4   Variable Delays

This section focuses on variable delays as illustrated in figure 11.5. Based on the previous discussion, an isolated single wave delay is examined. The overall system remains passive if this element stays passive, that is, if its output energy is limited by its input energy. And the position tracking is guaranteed if its output wave integral tracks its input integral. As such the forward and return delays may be different and are handled separately by duplicating the following efforts for both transmission paths.

11.4.1 Nonconservation of Variable Delays

First, understand the effect of a variable delay left untreated. Thus one uses

$$u_{out}(t) = u_{delay}(t) = u_{in}(t - T(t)) = u_{in}(t_{transmit}(t)) \qquad (11.14)$$

where $t_{transmit}(t)$ is the original transmit time of the data that are currently available at time t. The difference between the current time t and the corresponding transmit time $t_{transmit}(t)$ is the delay $T(t) = t - t_{transmit}(t)$.

As the delay varies, the wave signal is distorted. Indeed, if the delay increases, the transmit time increases only slowly and the input values are held longer. Hence the signal is stretched. In the extreme, if the transmit time stays constant and the delay grows as fast as time itself, the output becomes constant. Note one assumes that the order of the wave signal is preserved, that is, the data arrive at the remote site in the same order it is transmitted. This implies that the transmit time will never go backward and the delay time cannot increase faster than time itself,

$$\frac{d}{dt}t_{transmit}(t) \geq 0 \quad \Rightarrow \quad \frac{d}{dt}T(t) \leq 1 \qquad (11.15)$$

In contrast, if the delay shortens, the transmit time and hence the output signal change more rapidly. In essence the signal is compressed. Here the extreme case can lead to shock waves where data transmitted at different times arrive at the remote site simultaneously. This implies discontinuities and a jump in the output signal.

The delay variation and the corresponding changes in the wave signal may easily effect the system, if it is in any way correlated to the wave signal itself. For example, remember that the wave signal is interpreted as a push command. If the signal is expanded during a positive push and compressed during a negative pull command, the output will be biased in the position direction.

More formally, both the wave integral, which determines position tracking, and the wave energy, which determines passivity, are no longer conserved.

$$U_{out}(t) = \int_0^t u_{out}(\tau)d\tau \neq U_{in}(t - T(t)) \tag{11.16}$$

$$E_{out}(t) = \int_0^t \frac{1}{2}u_{out}^2(\tau)d\tau \neq E_{in}(t - T(t)) \tag{11.17}$$

Thus neither position tracking nor passivity are guaranteed. Also, as part of the signal compression and expansion, the frequency content changes which may produce other unexpected effects.

11.4.2 Integral Transmission

The above discussion illustrates that leaving variable delays untreated does not produce the desired results, as it does not preserve the wave integral or energy. As both of these quantities are central to the stability and performance of wave-based systems, the following solution is proposed, illustrated in figure 11.6.

Instead of transmitting the wave signal itself through the delay and then integrating, transmit both the wave integral and wave energy explicitly:

$$U_{delay}(t) = U_{in}(t - T(t)) = \int_0^{t-T(t)} u_{in}(\tau)d\tau \tag{11.18}$$

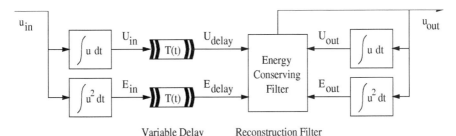

Figure 11.6: Transmitting the wave integral and energy, and then reconstructing the output wave signal based on these quantities can overcome the damaging effects of variable delay distortions.

$$E_{delay}(t) = E_{in}(t - T(t)) = \int_{0}^{t-T(t)} \frac{1}{2}u_{in}^2(\tau)d\tau \qquad (11.19)$$

The integration process thus remains consistent, though the resulting values are delayed. Also compute the equivalent quantities for the output wave signal $U_{out}(t)$ and $E_{out}(t)$ from (11.16) and (11.17). Then explicitly reconstruct the output wave signal such that its integral tracks the delayed input integral

$$U_{out}(t) \to U_{delay}(t) \qquad (11.20)$$

while using only the available energy

$$E_{out}(t) \le E_{delay}(t) \qquad (11.21)$$

The following section details this process.

Such an explicit reconstruction has several advantages. First, passivity is guaranteed by the definition of the system and is independent of the actual delay and or fluctuations thereof. One can build a passive and stable teleoperator on top of such communications. Second, explicit use of the wave integrals provides explicit position feedback. To this end, the wave integrals should be computed directly from position measurements.

If the delay is constant, no distortion is present and the output of such a reconstruction filter should equal the delayed original wave input. But should the delay fluctuate, the input may change more rapidly than can be matched with the incoming energy. In such cases, the filter is forced to smooth the signal to conserve energy. Much like wave-based controllers, one sees the system introduce an automatic performance limitation to remain passive, thereby providing a safety net against unpredictable fluctuations.

11.5   Reconstruction Filters

Many alternatives are possible to the filter problem defined by (11.20,11.21). To better understand the proposed solution, first examine the responses to impulse inputs. While the system is not linear, in the sense that the sum of two inputs does not necessarily produce the sum of the two individual outputs, such responses illustrate the basic behavior as a function of the system parameters. Also impulse inputs appear in real problems if transmission is temporarily blocked and the built-up data are released altogether.

11.5.1 Impulse Response

First define the wave distance to go $\Delta U(t)$ and energy reserve $\Delta E(t)$ available to the filter as

$$\Delta U(t) = U_{delay}(t) - U_{out}(t) \tag{11.22}$$

$$\Delta E(t) = E_{delay}(t) - E_{out}(t) \geq 0 \tag{11.23}$$

The solution takes the form

$$u_{out}(t) = \begin{cases} 2\alpha \dfrac{\Delta E(t)}{\Delta U(t)} & \text{if } \Delta U(t) \neq 0 \\[2mm] 0 & \text{if } \Delta U(t) = 0 \end{cases} \tag{11.24}$$

where α is a tunable parameter.

Such filters have several interesting properties. For zero input (holding $U_{delay}(t)$ and $E_{delay}(t)$ constant), they maintain the fixed ratio

$$\frac{\Delta E(t)}{\Delta U(t)^{\alpha}} = c \tag{11.25}$$

where c is a constant determined by initial conditions. Indeed,

$$\frac{d}{dt}\frac{\Delta E(t)}{\Delta U(t)^{\alpha}} = \frac{\Delta \dot{E} \cdot \Delta U - \alpha \cdot \Delta \dot{U} \cdot \Delta E}{\Delta U^{\alpha+1}}$$

$$= \frac{-\frac{1}{2}u_{out}^{2}\Delta U + \alpha \cdot u_{out}\Delta E}{\Delta U^{\alpha+1}} = 0 \qquad (11.26)$$

Thus for an impulse response, $u_{out}(t)$ may be written as

$$u_{out}(t) = \begin{cases} 2\alpha c \Delta U(t)^{\alpha-1} & \text{if} \quad \Delta U(t) \neq 0 \\ 0 & \text{if} \quad \Delta U(t) = 0 \end{cases} \qquad (11.27)$$

which is continuous at $\Delta U(t) = 0$ for $\alpha > 1$ and bounded for $\alpha = 1$. Furthermore one can write

$$\frac{d}{dt}\Delta U(t) = -u_{out}(t) = -2\alpha \frac{\Delta E(t)}{\Delta U(t)} = 2\alpha c \Delta U(t)^{\alpha-1} \qquad (11.28)$$

Consequently, both the distance to go $\Delta U(t)$ and the remaining energy $\Delta E(t)$ reach zero. Given the first order nature, $\Delta U(t)$ will not cross the origin during this convergence process and $\Delta E(t)$ remains positive.

Varying the parameter α selects how quick versus smooth the convergence process is. Figure 11.7 shows the impulse responses for the values of $\alpha = 1$, $\alpha = 1.5$, $\alpha = 2$. In the first case, the response is constant until the goal is reached and all energy has been used. This provides the fastest time to reach the goal, but it also contains discontinuities that may be disruptive in practice.

The response is linear for $\alpha = 1.5$ and exponential for $\alpha = 2$. The following developments will be constrained to the latter case. The impulse response is given by

$$u_{out}(t) = 4\frac{\Delta E_{0}}{\Delta U_{0}}e^{-\lambda t} \qquad (11.29)$$

where the bandwidth λ is determined as

$$\lambda = 4c = 4\frac{\Delta E(t)}{\Delta U(t)^{2}} \qquad (11.30)$$

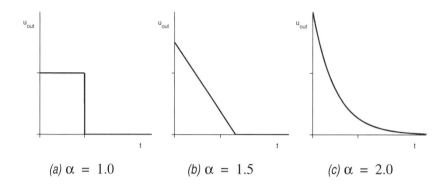

$$(a)\ \alpha\ =\ 1.0 \qquad\qquad (b)\ \alpha\ =\ 1.5 \qquad\qquad (c)\ \alpha\ =\ 2.0$$

Figure 11.7: Impulse responses for the reconstruction filter under a variety of parameter settings.

Notice that the speed of the response depends on the amount of energy $\Delta E(t)$ available for a given distance $\Delta U(t)$. The more energy, the faster the response. This also suggests a practical advantage: to limit the frequency content coming out of the reconstruction filter, one can saturate the stored energy reserve via

$$\Delta E(t) \le \frac{1}{4}\lambda_{max}\Delta U(t)^2 \tag{11.31}$$

11.5.2 Discrete Implementation

In practice, with a changing input $U_{delay}(t)$, the filter will not reach zero but continue to track the input with the first order behavior of (11.28). Limiting the energy according to (11.32) further smoothes the signal.

Nevertheless, when $\Delta U(t)$ and $\Delta E(t)$ approach zero the division in (11.24) is hard to compute. In addition, a finite sampling rate approximation may make $\Delta E(t)$ negative in violation of (11.21). Therefore, a discrete implementation is suggested, which accounts for a finite sample rate.

Rather than approximating the continuous derivatives, one rederives the equations starting with the requirement

$$\frac{\Delta E_{n+1}}{\Delta U_{n+1}^2} = \frac{\Delta E_n}{\Delta U_n^2} = c \tag{11.32}$$

for a zero input. This leads to the output

$$u_{out} = \begin{cases} \dfrac{4\Delta E \Delta U}{\Delta U^2 + 2\Delta E \Delta t} & \text{if } \Delta U^2 > 2\Delta E \Delta t \\[2em] \dfrac{\Delta U}{\Delta t} & \text{if } \Delta U^2 \leq 2\Delta E \Delta t \\[1em] 0 & \text{if } \quad \Delta U = 0 \end{cases} \qquad (11.33)$$

where Δt is the discrete sample time step.

Compared to the continuous time description, the first (main) case is adjusted to account for the zero order hold between samples. The second case is new and accounts for the possibility that the convergence happens within one time step. Indeed, in this case, explicitly reset $\Delta E = 0$ as the excess energy is unused. Finally, the third case applies if there is no further energy or distance to go.

11.6   Concluding Remarks

Teleoperation has traditionally been restricted to a few users of highly specialized equipment with dedicated transmission capabilities. The growth of the Internet is finally bringing this ability to physically interact at large distances to any PC user. It is also opening up the possibilities of multiple users interacting as a group, or sharing or coordinating control, or joining or retiring from a collaborative effort in an unregulated fashion. As these concepts are adapted to this general purpose network, new algorithms and approaches are needed to make systems robust and work side-by-side with e-mail, web browsing, and all other forms of information exchange found on the Internet.

Wave variables, described in patents [11, 12], are well suited for this application, in particular addressing the large fluctuations in delays that are inherent to the Internet. With minimal computation and no advance knowledge of the transmission characteristics, wave-variable encoding and filters enable network teleoperation, transparently during normal operation and degrading gracefully when the network is overloaded. Systems automatically feel soft and sluggish for large delays and become crisp for small delays. The developments may be combined with image transmission and audio streaming technologies to provide immersive teleoperation.

Finally, such teleoperative systems may also enable telesurgery through the

Internet and/or dedicated telephone lines (which have more stable delays). Early data appear to indicate that the added compliance introduced by the corresponding virtual tool would be negligible when interacting with soft tissues.

Acknowledgments

This report describes research done at the Nonlinear Systems Laboratory and the Artificial Intelligence Laboratory at the Massachusetts Institute of Technology. Support for preceding work has been provided in part by NASA/JPL. The authors are grateful to Ken Salisbury for providing the Phantom devices, and to Alain Carpentier for stimulating discussions on telesurgery.

References

[1]   Anderson, R. J., and Spong, M.W. (1989). Bilateral control of teleoperators with time delay. *IEEE Transactions on Automatic Control* 34 (5): 494–501.

[2]   Backes, P. G., Tso, K. S. and, Tharp, G. K. (1998). Mars Pathfinder Mission Internet-Based Operations Using WITS. In *Proc. of 1998 IEEE Int. Conf. on Robotics and Automation*.

[3]   Lohmiller, W. and Slotine, J.-J. E. (2000). Control System Design for Mechanical Systems Using Contraction Theory. *IEEE Transactions on Automatic Control* 45 (4).

[4]   Niemeyer, G., and Slotine, J.-J. E. (1991). Stable adaptive teleoperation. *IEEE Journal of Oceanographic Engineering*, 16 (1): 152–162

[5]   Niemeyer, G., and Slotine, J.-J. E. (1997a). Designing force reflecting teleoperators with large time delays to appear as virtual tools. In *Proc. of 1997 IEEE Int. Conf. on Robotics and Automation*, pp. 2212–2218.

[6]   Niemeyer, G., and Slotine, J.-J. E. (1997b). Using wave variables for system analysis and robot control. In *Proc. of 1997 IEEE Int. Conf. on Robotics and Automation*, pp. 1619–625.

[7]   Niemeyer, G., and Slotine, J.-J. E. (1998). Towards Force-Reflecting Teleoperation Over the Internet. In *Proc. of 1998 IEEE Int. Conf. on Robotics and Automation*.

[8]   Sheridan, T. B. (1989). Telerobotics. *Automatica*, 25 (4): 487–507

[9]   Sheridan, T. B. (1992). *Telerobotics, Automation, and Human Supervisory Control*. Cambridge: MIT Press.

[10] Slotine, J.-J. E., and Li, W. (1991). *Applied Nonlinear Control*. Englewood Cliffs, NJ: Prentice Hall.

[11] Slotine, J.-J. E., and Niemeyer, G. (1993). Telerobotic System. U.S. Patent

no. 5266875.

[12] Niemeyer, G., and Slotine, J.-J. E. (2000).Teleoperation with variable delay. U.S. Patent no. 6144884

12 VISIT: A Teleoperation System via the Computer Network

Kazuhiro Kosuge, Jun Kikuchi, and Koji Takeo

12.1   Introduction

With the development of information infrastructure, teleoperations virtually from any place at any time in the world have become possible by utilizing the computer network. ISDN, cellular phone systems, and other systems are available utilizing the computer network. The intelligent mechanical system, such as a robot is only one device, which transforms information into physical energy to execute a physical task in the real world. The development of a robot system applicable to a variety of tasks is one of the key issues relating to further possibilities of the information infrastructure.

This chapter considers issues about teleoperation via the computer network. First, the issue of variable communication time delay is discussed. The virtual time-delay method will be introduced for manual teleoperation with bilateral feedback. Then the use of predictive display for the manipulation of dynamic environment under communication time delay is discussed. Several simulations will illustrate the effectiveness of the prediction for the problem. Recent motion-control technology, computer vision technology, and a graphic-user interface are integrated and then a teleoperation system with a visual interface system for interactive task executions, which is referred to as VISIT is proposed. Experiments using a prototype system will illustrate the concept.

12.2   Design

12.2.1 Software Architecture and Design

Teleoperation System with Communication Time Delay

How to increase the reality of teleoperation has been a big obstacle for manually operated teleoperation systems. Bilateral feedback is one of the effective schemes for hi-fi teleoperation, with which force information is fed back to a master from a slave. A bilateral teleoperation system has the problem of relating to the

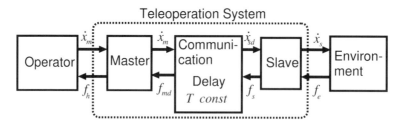

Figure 12.1:   Structure of traditional teleoperation system.

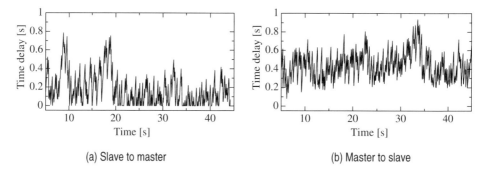

(a) Slave to master   (b) Master to slave

Figure 12.2:   Time delay on computer network.

communication time delay between a master and a slave (figure 12.1). When velocity and force information are transmitted via the communication block with time delay, even a small time delay easily destabilizes the bilateral teleoperation system.

Anderson and Spong suggested a method of overcoming this instability by using scattering transformation [1]. In this method, the communication block is assumed to have a constant time delay in both directions between a master and a slave. The scattering transformation guarantees the passivity of the communication block and ensures the passivity of the total system. The passive teleoperation system is known to be stable for the manipulation of unknown environment [1].

When a communication block is composed of a computer network, time delay is not constant as shown in figure 12.2. This shows an example of the time delay between two workstations, at a master site and at a slave site, connected to a network. Data are transmitted by a stream socket via Ethernet. The size of the

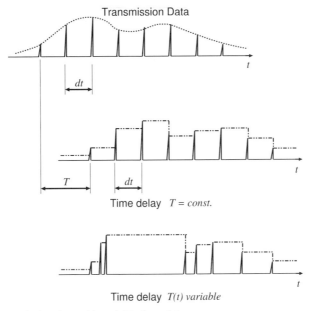

Figure 12.3: Transmission data with variable time delay.

datum was 10 bytes and the datum was sent every 2ms. This shows that time delay was up to 0.94s and irregularly changed.

If the communication time delay is constant and the data are sent with a constant time interval from a master site, the data are received with the same changes. Even if the data were sent with a constant time interval, the data are received with variable time interval that irregularly changes as illustrated in figure 12.3. Multiple data may be received for a short time and data may not be received for a long time.

Suppose that a communication block has a delay which changes from T to $T-dT$, as illustrated in figure 12.4. If wave-form data are transmitted via this communication block, the output wave-form changes as shown in figure 12.4. Namely, when the time delay is constant, the communication block adds only phase lag to the signal, but when the time delay changes, even the frequency of the signal changes.

The communication block with variable time delay is a time varying system; therefore, passivity of the communication block with only scattering transformation cannot be guaranteed.

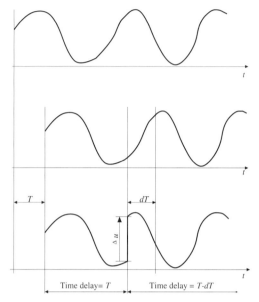

Figure 12.4: Wave form with variable time delay.

Virtual Time-Delay Method

This section presents the virtual time-delay method to overcome the problem of a variable time delay. This method keeps the apparent time delay constant and keeps the stability of the system using the scattering transformation even with the variable transmission delay.

First, the traffic of the network is observed and the maximum time delay of the transmission, T_c, between a master site and a slave site is decided. Velocity and force information, which are sampled with constant sampling time intervals are sent to the other site together with its sampling time T_i as illustrated in figure 12.5.

The data received after the time delay T_i, which may be shorter than the maximum time delay T_c, are held in a receiver until the delay reaches to the maximum time delay T_c. Thus, T_c is the virtual time delay of the system. Since this method makes the apparent time delay of the system equal to the virtual time delay, which is constant, then the passivity of the transmission system is guaranteed by conventional scattering transformation. The virtual time delay can be selected as long as it is larger than the maximum time delay; however, the selection of large time delay decreases the maneuverability of the system. The new

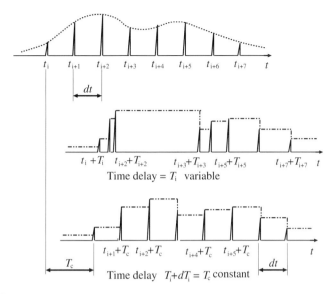

Figure 12.5: Compensation to keep time delay constant.

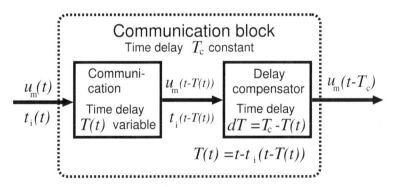

Figure 12.6: Structure of compensation system.

compensation virtual time delay method described above (see also figure 12.6 and 12.7) was implemented in a single axis master slave-manipulator.

The manipulator is controlled to have impedance parameters as follows:

$$M\ddot{x}_m + D\dot{x}_m = f_h - f_{md} \tag{12.1}$$

$$M\ddot{x}_s + D\dot{x}_s = f_s - f_e \tag{12.2}$$

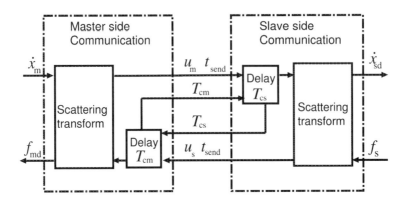

Figure 12.7: Communication block with compensation system.

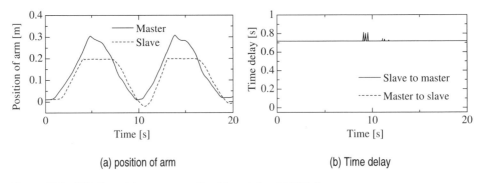

(a) position of arm

(b) Time delay

Figure 12.8: With time delay compensation (time-delay: 0.718 [s]).

where, x_m and x_s is the position of the master arm and the slave arm. f_s is the force information transmitted to the master side. The operator force f_h and the environment force f_e are acquired via force sensors attached to a hand of each manipulator. Actual manipulators are controlled to follow motion of the above impedance model applied by those forces. Figure 12.8 shows an example of the experimental results. The detailed experimental results are shown in [8].

12.2.2 Interface Design

Teleoperation with Predictive Display

In general, visual information from the remote site is provided to the operator for the improvement of safety and efficiency of operation; however, the operator is

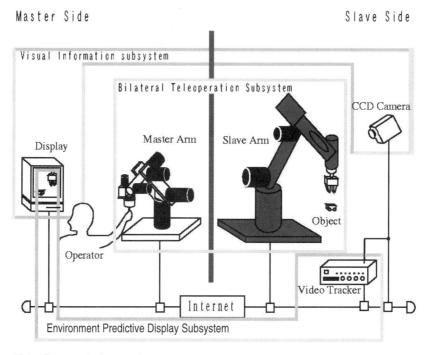

Figure 12.9: Proposed teleoperation system.

able to recognize only past situations of the remote site because visual information has varying communication time delay. If it is possible to provide pictures of the predictions of the slave manipulator and the environment to the operator, the operator can execute tasks virtually at real time and the efficiency of the operation improves.

Teleoperation systems with predictive display, which were introduced in this chapter consist of the following three subsystems (figure 12.9):

- Bilateral teleoperation subsystem.

- Visual information subsystem.

- Environment predictive display subsystem.

Each subsystem is connected to a computer network and exchanges information with each other over the network. For the bilateral teleoperation subsystem, the virtual time-delay method explained in the previous section as implemented. The following section illustrates the visual information subsystem and the environment predictive display subsystem.

Visual Information Subsystem

The visual information subsystem transmits the pictures about a task from the remote site to the operator's site. At the remote site, an image of the slave manipulator and its environment are captured into a computer system and sent to a computer system in the operator's site via the computer network at each image sampling time. The operator can handle the master manipulator watching the image transmitted and shown in the computer display at the operator's site.

Environment Predictive Display Subsystem

As mentioned above, the visual information subsystem as designed so that information is sent over the computer network; visual information provided by the visual information subsystem has a certain amount of time delay. Since this delayed information is usually less effective for the operator to execute a task, real-time information was considered by estimating the remote site environment.

Consider the case of the manipulation of an object. Here the motion of the object using the Kalman filter and so forth can be estimated with the measurements of the states of the object, such as position, velocity and so on, even if the dynamics of the object is unknown. The motion of the object after a certain time period can be predicted using the equation of motion about an appropriate model of an object.

The information of the predicted position is transmitted to the computer at the operator's site via the computer network with the varying communication time delay. To take this time delay into account, a time stamp is attached to each data block transferred to the master site. At the master site, according to the appended time stamp, the data stream is regulated to have constant time delay and the information is provided to the operator.

12.3   The Setup: VISIT

12.3.1   General Description

The teleoperation system referred to as VISIT, which is proposed in this chapter, has the following features:

- The robot and the teleoperation system are connected via available information infrastructure such as the Internet, ISDN, cellular phone, and so forth so that the system is available from any place at any time.

- The system is designed based on advanced robotic technologies, such as computer vision, advanced motion control and so forth so that the operator can execute a task in a remote site by utilizing GUI and a pointing device.

- The skill required to execute a task is prepared as an ITEM, an interactive task execution module. The operator execute a task interactively with the system using an appropriate ITEM.

- The system is portable and can be applied to not only the teleoperation but also the teach-in operation of the robot.

The system includes the following modules.

Graphic-User Interface Module

The graphic-user interface module enables an operator to give a command to a robot in a remote site using a pointing device such as a mouse. The real vision is superimposed on the computer graphics through this module. The robot motion simulator will be included in the module.

Computer Vision Module

The computer vision module is used to identify which object the operator would like to manipulate. By this module, the operator could indicate the object using a pointing device. It is assumed that the system has a knowledge database of the objects involved in the task.

Interactive Task Execution Module (ITEM)

The ITEM enables the operator to execute a task, which requires human skill and dexterity, by giving commands interactively to the robot using the graphic-user interface and a pointing device. The ITEM is in charge of giving commands to one of the motion-control modules of the robot so as to execute the task interactively. Several classes of ITEMs are prepared for this purpose.

Motion-Control Module

Advanced motion-control schemes, such as trajectory tracking control, stiffness control, damping control, impedance control hybrid, and position-force control, are provided by this module. The time delay problem is also taken into account by this module and the ITEM is designed with this module.

Communication Module

The communication module provides protocol-hardware independent communication channel between modules of the VISIT. Each module of the VISIT is designed to communicate via this module for portability and expandability.

12.3.2 Experience

A prototype of VISIT is being developed for an industrial robot manipulator. Figure 12.10 left shows the GUI system build on top of the X window system. The parameters for the image processing, as well as the command to the robot, are displaycd and could be modified if necessary.

The sequence of the task execution by a prototype system is shown as follows:

Object Identification

First, the object to be manipulated should be identified by the system. The computer vision module measures the shape and the position of the object in the area of interest that is specified by the operator (indicated by the bright rectangle on the live video image in figure 12.10 right). The objects are shown on the sub window as wire-frame graphics (small window below the live video window in figure 12.10 right). If the result is not satisfactory, measurements will be done repeatedly by changing some parameters.

Task specification

After the object is identified, a 3-D model of the object is superimposed on the live video image. The task is specified by moving the 3-D model on the live video image in a "drag and drop" manner (figure 12.11 left). The precise position and orientation of the object is also provided on the floating palette with sliders for fine adjustment.

Task execution

Finally, the task execution command is issued to the remote manipulator. The motion of the remote manipulator could be monitored on the live video window (figure 12.11 right).

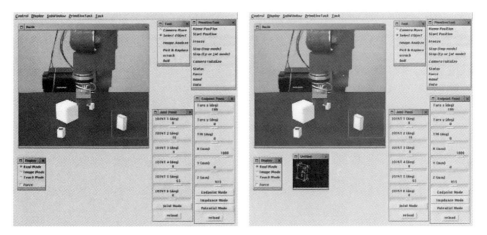

Figure 12.10: Left: GUI of VISIT prototype system.
Right: Object identification.

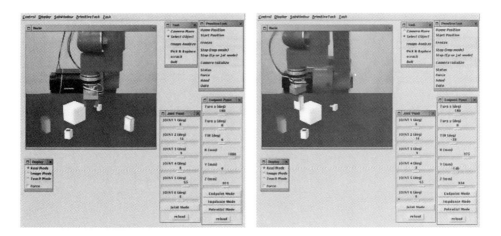

Figure 12.11: Left: Task specification on the GUI system.
Right: Robot moving an object by VISIT.

12.4   Conclusions

Issues and their solutions relating to teleoperation via computer network were discussed, then VISIT, a teleoperation system with a visual interface system for interactive task execution was proposed. A prototype system of VISIT was introduced to illustrate the concept.

References

[1] Robert J. Anderson and Mark W. Spong. Bilateral Control of Telemanipulators with Time Delay. *IEEE Transactions on Automatic Control*, vol. 34, no. 5, 1989, pp. 494–501.

[2] Y. Yokokoji and T. Yoshikawa. Bilateral Control of Master-Slave Manipulators for ideal kinesthetic Coupling – Formulation and Experiment. *IEEE Transactions on Robotics and Automation*, vol. 10, no. 5, 1994, pp. 605–619.

[3] K. Furuta and K. Kosuge. Master Slave Manipulator based on Virtual Internal Model Following Control Concept. *IEEE International Conference on Robotics and Automation*, 1987.

[4] J. Dudragne, C. Andriot, R. Fournier, and J. Vuillemey. A Generalized Bilateral Control Applied to Master-Slave Manipulators. *Proceedings of 20th ISIR*, 1988, pp. 435–442.

[5] Robert J. Anderson. Building a Modular Control System using Passivity and Scattering Theory. *IEEE International Conference on Robotics and Automation*, 1996, pp. 698–705.

[6] K. Kosuge, T. Itoh, and T. Fukuda. Scaled Telemanipulation with Time Delay. *IEEE International Conference on Robotics and Automation*, 1996, pp. 2019–2024.

[7] G. Hirzinger, K. Landzettel, and C. Fagerer. Telerobotics with Large Time Delays — the ROTEX experience. *IEEE/RSJ/GI International Conference on Intelligent Robots and Systems*, 1994.

[8] K. Kosuge, H. Murayama, and K. Takeo. Bilateral Feedback Control of Telemanipulators via Computer Network. *IEEE International Conference on Intelligent Robots and Systems*, 1996.

[9] J. Kikuchi, K. Takeo, and K. Kosuge. Teleoperation System via Computer Network for Dynamic Environment, *IEEE International Conference on Intelligent Robots and Systems*, 1998

13 Internet-based Ground Operations for Mars Lander and Rover Missions

Paul G. Backes, Kam S. Tso, Jeffrey S. Norris, and Gregory K. Tharp

13.1 Introduction

The Web Interface for Telescience (WITS) provides Internet-based ground operations for planetary lander and rover missions. WITS provides an integrated environment for operations at the central operations location, collaboration by geographically distributed scientists, and public outreach. Collaboration by geographically distributed scientists at their home institutions enables participation in missions by a greater number of scientists and reduces operations costs. Providing the same operations tool, with real mission data, to the public enables the public to have a better understanding of the mission and a more engaging mission experience. An initial version of WITS provided some of the features needed for a planetary rover mission [1]. Based upon experience with WITS as an evaluation and public outreach tool in the 1997 Mars Pathfinder mission [2] and rover field tests [3, 4], WITS was redesigned and reimplemented to provide the features needed for planetary mission operations. This chapter describes the new WITS system.

Through the use of WITS, the Mars Polar Lander (MPL) mission was the first planetary mission to utilize Internet-based ground operations to enable geographically distributed scientists to collaborate in daily mission command sequence generation. The same WITS system was used for the MPL mission and for commanding the FIDO rover. Other examples of Internet-based robot operation can be found in [5, 6].

13.1.1 Mars Polar Lander Mission

The Mars Polar Lander landed near the south pole of Mars on December 3, 1999, and was to perform an approximately three-month mission [7]. Unfortunately, communication with the lander was never achieved so commanding the lander was not possible. Results shown in this chapter are from commanding the lander in the University of California at Los Angeles (UCLA) MPL testbed. An artist's drawing of the lander is shown in figure 13.1. The lander carried the Mars Volatiles and

Figure 13.1: Artist's drawing of Mars Polar Lander on Mars.

Climate Surveyor (MVACS) instrument suite. The mission operations center was at UCLA.

WITS served multiple purposes for the MPL mission. It was the primary operations tool for visualization of downlink data and generation of command sequences for the robotic arm (RA) and robotic arm camera (RAC). It was also used as a secondary tool for command sequence generation for the lander mast-mounted stereo surface imager (SSI) stereo camera, for example, for visualizing footprints on the surface where SSI images would be taken. WITS also enabled Internet-based users to generate command sequences for the RA, RAC, and SSI. For example, scientists at the University of Arizona, who were responsible for the RAC and SSI, were able to generate sequence inputs from Arizona. This capability would enable them to participate from Arizona for part of the mission. Also, a separate WITS system was provided to the general public to download to their home computers thus enabling them to plan and simulate their own missions.

13.1.2 FIDO Rover Operations

WITS was used to command the Field Integration Design and Operations (FIDO) rover in the two week field test at Silver Lake in the California Mojave desert in April 1999. The FIDO rover has been developed as a prototype vehicle for testing new planetary rover technologies and testing operations scenarios for the Mars '03

Figure 13.2: FIDO rover.

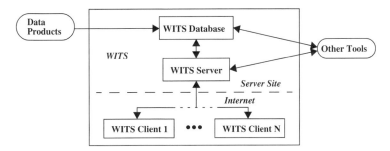

Figure 13.3: WITS architecture.

Athena rover mission [8, 9]. The FIDO rover is shown in figure 13.2. WITS supported operations in the field as well as remote operations over the Internet.

13.2   System Description

The WITS architecture is shown in figure 13.3. The WITS database holds downlink data products and uplink sequence information. The WITS server provides communication between the WITS database and the WITS clients. The WITS clients are distributed over the Internet and provide the interface to the users to view downlink data and generate command sequences. Internet security was integrated between the server and clients. Communication between the client and server is implemented using Java remote method invocation (RMI). The WITS database is a structured file system. Other tools, for example, planetary ephemeris tools or other sequence tools, can interact with WITS by reading and writing to the WITS database or by direct communication with the WITS server.

The WITS client and server are implemented using the Java2 programming language, including the Java3-D and Java cryptography extensions. The client is run as a Java applet in a web browser, or in an appletviewer, or as a Java application. Users must first download the Java run-time environment and Java3-D.

13.2.1 Internet Security

A critical element in Internet-based mission operations is Internet security. Each day, large amounts of data are received from the spacecraft at the mission operations center and placed into a local database for processing and viewing by mission scientists. To enable collaboration in daily sequence generation by distant, Internet-based scientists, a secure and efficient way to deliver the data to the remote scientists as well as receive inputs from them is needed. The WITS Encrypted Data Delivery System (WEDDS) was created to provide the required secure Internet-based communication for WITS. WEDDS was integrated with WITS for the MPL mission, but it is designed to work with any mission application with little modification. WEDDS operates in a fashion that is transparent to the remote user. Files simply appear on the remote user's machine as they become available, and connections are made securely without any additional effort on the part of the user.

All WEDDS connections are authenticated using the NASA Public Key Infrastructure (PKI) [10]. After authentication, all WEDDS communications are made through secure sockets layer (SSL) sockets and are encrypted using the Triple-DES-EDE3 algorithm [11,12]. The following points are some of the advantages of WEDDS for Internet-based data distribution: (1) WEDDS does not require the remote users to request a data update. The data is automatically delivered as it becomes available. (2) The WEDDS administrator can specify on a user-by-user basis exactly which users will receive a particular type of file or directory. (3) Since WEDDS is written entirely in Java, it can run on most computers without modification. (4) WEDDS provides a high level of data

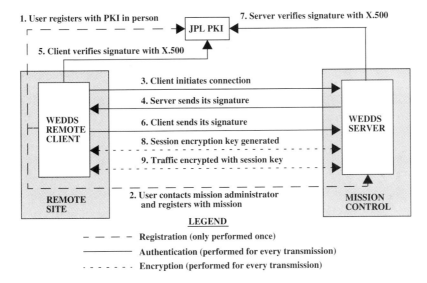

Figure 13.4:   Secure Internet communication steps.

security by using the SSL algorithm to perform authentication and the Triple-DES-EDE3 algorithm for encryption. The SSL protocol protects nearly every online commerce transaction, and breaking the Triple-DES-EDE3 encryption algorithm, even if every computer in the world was used, would take millions of years [13]. (5) WEDDS clients can be allowed to transfer files back to the mission control center. Files from users are stored on the server in a compressed, enveloped form that allows them to be scanned for hostile code. Since every client is authenticated using the NASA PKI, unauthorized users can not transmit files to the server.

WEDDS is implemented as two Java programs, a server and a client, using the publicly available Entrust Java Toolkit for its low-level authentication and encryption tasks [14]. For each mission, there is typically one server, operating behind a mission firewall, and many clients, one on each remote user's machine.

Figure 13.4 illustrates the steps necessary for a single WEDDS transaction. Steps 1 and 2 occur once, before the beginning of the mission, while steps 3 through 9 occur for every transaction. In step 1, a remote user must obtain a security profile from the NASA PKI, which requires appearing in person at a NASA center security office. A security profile is a set of files on a floppy disk, protected by a password, that contain a user's private key. Users need individual private key to positively identify themselves online. The WITS server is also issued a security profile so that it can prove its identity to remote users. In step 2, users must contact the mission administrator and request that their profile be given access to mission data.

Steps 3 through 9 are repeated for every transmission from the client to the server or from the server to the client. In steps 4 through 7, the server and client exchange digital "signatures" generated from their security profile. They verify these signatures by communicating with the NASA PKI. This process, SSL authentication, relies on the fact that it is nearly impossible for someone to generate another user's digital signature without that user's private key and password. In step 8, the last step in establishing an SSL connection, the client and server use the Diffie-Hellman key agreement protocol [15] to generate a unique symmetric encryption key that will be used to encrypt the traffic for this transaction. This encryption key is never sent as clear-text over the Internet, and all subsequent traffic for this transaction is encrypted, making WEDDS transactions invulnerable to eavesdropping or "packet-sniffing" attacks. In addition, every WEDDS transaction is protected by a new encryption key.

13.2.2 Distributed Collaboration

Internet-based collaboration is achieved in WITS by providing Internet-based users with daily downlink data updates and allowing them to specify targets and generate command sequences and save them to the common server. The Internet-based users can access targets and sequences created by other users and can use targets from the other users' sequences. The sequence window file pull-down menu enables sequences to be loaded from the common server (e.g., located at UCLA), or to be saved to the common server. Users can also save and load sequences to their local computers.

13.2.3 Downlink Data Visualization

Downlink data from the lander or rover are provided to the user via various views (figure 13.5). Example views from the MPL mission are used here. The same views, but with rover data, are used for FIDO rover commanding. Two windows provide the user with available data to be visualized. The results tree window (not shown in the figure) displays the available downlink data for the mission by date of downlink. The plan window (figure 13.5) displays available views for a specific

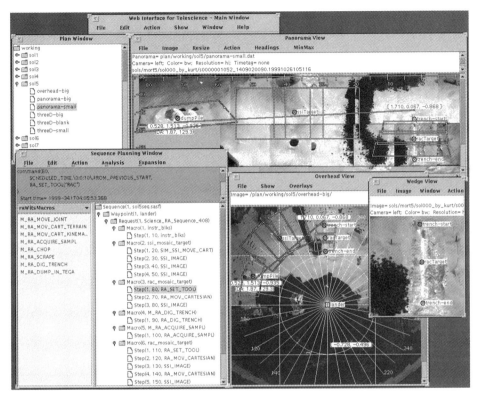

Figure 13.5: Panorama, wedge, and overhead views and sequence and plan windows.

plan (a plan includes view definitions and sequence information for generating one uplink sequence). Each of the specific views has a specified set of downlink data it uses, so definitions of these views may be updated each day and put in the new plan. The user opens a view to visualize downlink data by clicking on the item. The various types of views are described below.

The descent view (not shown in the figure) provides images taken from the spacecraft during descent to the surface and shows the landing location. For rover missions, the rover traverse path during the mission is shown. The overhead view (figure 13.5) shows the immediate area around the lander or rover from above. Targets are displayed in the overhead view as well as selected points and view objects. Clicking on a point in the overhead view causes the x,y position to be displayed at the clicked location. Clicking and dragging causes a ruler to be shown with the start and end points and distance displayed.

The panorama view (figure 13.5) is a mosaic of images taken by a stereo camera (on a mast on the lander in the MPL mission and on a deployable mast on the FIDO rover). Selecting a point in an image causes the x, y, z position at that point on the surface to be displayed. The point can be turned into a known science target via a menu option. The panorama view can be shown in 1/4, 1/2, and full scale. The view can also be shown in anaglyph stereo via a menu option. Clicking and dragging causes a ruler to be displayed with the start and end points x, y, z values and the distance and azimuth between the points. The wedge view displays one image with various viewing options.

The contrast adjuster view (opened from a wedge view pull-down menu) enables the contrast to be adjusted for a wedge view image. The minimum and maximum desired pixel intensities are selected via scroll bars and then the pixel intensity values of the image are linearly stretched to have the selected pixel intensities become minimum (0) and maximum (255). The histogram of the initial and adjusted images are also shown.

The 3-D view, shown in figure 13.6, provides a 3-D solid model visualization. Sequence simulation and state is visualized in the 3-D view.

13.2.4 Sequence Generation

The views discussed above provide a means for visualizing downlink mission data. WITS also provides various windows and features for command sequence generation. WITS enables 3-D locations to be turned into known targets to be used as parameters in commands. Targets are displayed in the various views as pink circles.

The sequence window, shown in figure 13.5, is used to generate a command sequence. A command sequence has a hierarchy of elements. The hierarchy, in descending order, is: sequence, waypoint, request, macro, and step. A request represents a high-level task. A macro is the functional element in WITS by which the user specifies commands and parameters. Macros have expansions into steps. A step is a low-level command that will be uplinked to the spacecraft. WITS can generate various format output sequences. The sequence window shows the sequences in one plan. Multiple sequences can be displayed. A plan generally represents the planning elements to generate one command sequence to be uplinked to the lander or rover. The sequences are shown on the right-hand side of

Figure 13.6: 3-D view with lander and terrain visualization.

the sequence window. Supporting multiple sequences is useful for integration of subsequences from different scientists or subsequences for different instruments into the final uplink sequence. There are various sequence editing features in the sequence window.

A list of macros that can be inserted into a sequence is shown on the left side of the sequence window. Multiple lists of macros are available; choosing between macro lists is done via the pull-down menu above the macro list. A macro is inserted into a sequence by selecting the location in the sequence for it to be inserted and then double clicking on the macro in the macro list. Double clicking on a macro in the sequence causes the macro window to pop up (figure 13.7), where parameters for that macro are specified. A macro can generate view objects that are displayed in the views to indicate what the macro is producing. Figure 13.5 shows square outlines that are view objects for SSI imaging commands. They represent where images will be taken by the SSI in the current sequence. Above the dump pile is a view object of a RAC image and above the trench is a view object of a

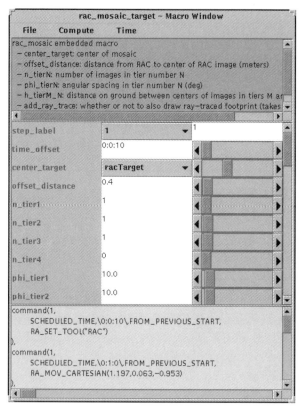

Figure 13.7: Macro window.

three-image RAC mosaic. View objects can also be generated by parsing the steps directly.

Resource analysis and rules checking are provided for sequences. The duration, energy, and data volume for each step of the sequence are computed and stored along with the cumulative duration, energy, and data volume at each step. Sequences are also automatically checked to verify that they satisfy specified sequence rules.

The sequence state at a step is displayed when a user clicks on the step in the sequence window. Step expansion and resource information is displayed at the top of the sequence window and the system state is visualized in the 3-D view.

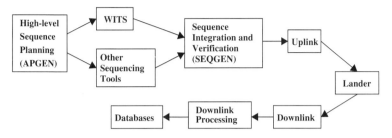

Figure 13.8: MPL mission operations architecture.

13.3   Performance

13.3.1 Mars Polar Lander Mission Results

WITS was a part of the complete MPL mission ground operations system. A simplified diagram of the MPL ground operations system is shown in figure 13.8. Downlink data from Mars was to be processed and put in databases, for example, the WITS database. Sequence generation began by using a sequencing tool called APGEN which generated daily high-level sequences for all the lander instruments. The sequences for the different instruments were sent to sequence generation systems specific to each instrument. Included in the sequences were requests that included text descriptions of what needed to be done in each request and how much time, energy, and data volume was allocated for each request. WITS was the sequence generation system for the RA and RAC and some SSI operations. WITS generated the low-level commands to achieve the goals specified in the requests and within the resource allocations specified. The multiple sequencing systems then output their sequences to the SEQGEN planning tool where all the low-level sequences were integrated and resource checking on the integrated sequence was done and final sequence modifications were made to ensure a valid sequence within resource constraints. The final sequence was then sent into the uplink process, eventually to be received at the lander.

An important motivation for the development and use of WITS in a planetary mission is its use in public outreach. A separate version of WITS was made available to the general public to download and run on their home computers. A subset of mission data was placed in the public WITS database. Since the actual mission commands and flight software are proprietary, new arm kinematics and new commands were used in the public version. The public was then able to view mission data and plan and simulate their own missions. The site to download the public MPL mission WITS can be found at URL http://robotics.jpl.nasa.gov/tasks/wits/. The public outreach site was hosted at Graham Technology Solutions, Inc. The KEWi Internet help desk system from 3Si Inc. [16] was used to provide technical support to the public (it is also used for FIDO rover WITS technical support). With KEWi, public users can query for information about WITS, using the KEWi search engine, or submit service tickets that the KEWi WITS support staff then reply to.

13.3.2 FIDO Rover Field Test Results

A science and engineering team used WITS locally in the field to command the FIDO rover in the April 1999 field test at Silver Lake, California. Geographically distributed high school students commanded the FIDO rover during the last two days of the field test using WITS via the Internet (connected to the field via a satellite connection). High schools from Los Angeles, CA, Phoenix, AZ, Ithaca, NY, and St. Louis, MO, made up the LAPIS student tests mission team. The students viewed the downlink data and generated the command sequences. After the students finished generating a sequence, they submitted it to the common database and the field located operators reviewed it and sent it for execution on the rover. During field test operations, WITS had the functionality to command eight different instruments (pancam, navcam, hazcams, bellycam, IPS, color microscopic imager, moessbauer, and minicorer) as well as the mast, arm, and rover motions. WITS also had the functionality to display the downlink data results for all instruments in graphical and/or textural format.

13.4   Conclusions and Future Work

WITS provides a new Internet-based operations paradigm for planetary mission operations. WITS was used in the Mars Polar Lander mission ground operations for downlink data visualization and command sequence generation for the robotic arm and robotic arm camera and as a secondary tool for the stereo surface imager. With the use of WITS, the MPL mission was the first planetary mission to utilize Internet-based ground operations. WITS can also provide Internet-based operations for future Mars rover missions, as demonstrated in its use in the FIDO rover field test. The integrated visualization, sequence planning, and Internet security features of WITS make it a valuable tool for planetary mission operations. WITS is also an engaging public outreach tool that the public can use to visualize mission downlink data and plan and visualize their own missions.

WITS was also prepared for the FIDO rover field test in May 2000. The primary focus of current development is to enhance the visualization and simulation capabilities, for example, simulation of the rover traverse and simulation of rover instrument arm and minicorer deployment against targets.

Acknowledgments

The research described in this chapter was carried out by the Jet Propulsion Laboratory, California Institute of Technology, under a contract with the National Aeronautics and Space Administration. The contribution by IA Tech, Inc. was supported by a contract from the NASA Small Business Innovation Research program.

References

[1] P. Backes, G. Tharp, and K. Tso. The Web Interface for Telescience (WITS). *IEEE International Conference on Robotics and Automation,* p. 411 – 417, Albuquerque, New Mexico, April 1997.

[2] P. Backes, K. Tso, and G. Tharp: Mars Pathfinder Mission Internet-Based Operations Using WITS, *IEEE International Conference on Robotics and Automation*, p. 284 – 291, Leuven, Belgium, 1998.

[3] S. Hayati, R. Volpe, P. Backes, J. Balaram, R. Welch, R. Ivlev, G. Tharp, S. Peters, T. Ohm, and R. Petras: The Rocky7 Rover: A Mars Sciencecraft Prototype. *IEEE International Conference on Robotics and Automation,* p. 2458 – 2464, Albuquerque, New Mexico, April 1997.

[4] R. Volpe. Navigation Results from Desert Field Tests of the Rocky7 Mars Rover Prototype. *International Journal of Robotics Research,* special issue on field and service robots, 18 (7), July 1999.

[5] E. Paulos and J. Canny. Delivering Real Reality to the World Wide Web Via Telerobotics. *IEEE International Conference on Robotics and Automation,* p. 1694 – 1699, Minneapolis, Minnesota, April 1996.

[6] K. Goldberg, M. Mascha, S. Genter, N. Rosenberg, C. Sutter, and J. Wiegley. Robot Teleoperation Via WWW. *IEEE International Conference on Robotics and Automation.* May 1995.

[7]   NASA/UCLA: Mars Surveyor 98 Lander, http://mars.jpl.nasa.gov/msp98/lander/, 1999.

[8]   P. Schenker et al. New Planetary Rovers for Long Range Mars Science and Sample Return. *Intelligent Robotics and Computer Vision XVII*, SPIE Proc. 3522, November 1998.

[9]   R. Arvidson et al. FIDO: Field-Test Rover for 2003 and 2005 Mars Sample Return Missions. *30th Lunar and Planetary Science Conference, Houston*, Texas, March 15 – 19, 1999.

[10] R. Rivest, A. Shamir, and L. Adleman. A Method for Obtaining Digital Signatures and Public-Key Cryptosystems. *Communications of the ACM* 21 (2), p. 120 – 126, February 1978.

[11] SSL 3.0 Spec. http://home.netscape.com/eng/ssl3/, Netscape Corporation.

[12] RSA Labs FAQ, Section 3.2. DES, http://www.rsasecurity.com/rsalabs/faq/3-2.html, RSA Security Incorporated, 1999.

[13] M. Wiener. Performance Comparison of Public-Key Cryptosystems. *CryptoBytes* 3 (3), p. 1 – 5, 1998.

[14] Entrust/Toolkit Java Edition, http://developer.entrust.com/java/, Entrust Technologies, 1999.

[15]   W. Diffie and M. Hellman. New Directions in Cryptography. *IEEE Transactions on Information Theory*, IT-22, p. 644 – 654, 1976.

[16]   3Si, Inc.: http://www.3si.com/.

14 Robust Visualization for Online Control of Mobile Robots

Wolfram Burgard and Dirk Schulz

14.1 Introduction

The Internet provides a unique opportunity to teleoperate and monitor mobile robots. Online controlled mobile robots can give people all over the world the ability to become telepresent at distant places. Internet-based tele-experimentation systems for mobile robots also give distributed research groups located at distant places the ability to carry out joint experiments. In this way, they can share expensive robot platforms and furthermore save travel expenses. Finally, the Internet can be used for online demonstrations with mobile robots, for example, during the presentation of research results at conferences.

All these applications of web interfaces for mobile robots require accurate visualization techniques, including a high level of detail as well as high update rates. Unfortunately, the Internet only provides a restricted bandwidth and arbitrarily large transmission gaps can occur. Because of the limited bandwidth, video streams with appropriate resolutions do not achieve the frame rate required for a smooth visualization of the robot's actions. Installing static monitoring cameras, a method frequently used on the Internet, is also not a feasible approach; they only provide a reduced field of view and important details may also be occluded.

This chapter presents the *robotic tele lab* system (short RTL-system) that provides accurate and smooth real-time 3-D visualizations of the movements of an autonomous mobile robot over the Internet. The visualization component uses a 3-D model of the robot and its environment to display the current state of the robot. Given this model, only the current state of the robot needs to be transferred over the Internet to animate the robot's actions. As this is only a small amount of information, these updates can be carried out at high frequencies. On the Internet, however, several of the state updates may get lost. To bridge the resulting transmission gaps, the visualization component of the RTL-system also performs predictive simulations. In this way, it is able to accurately visualize the state of the robot even during transmission gaps of several seconds.

The RTL-system is designed for mobile robots operating in dynamic environments where the position and orientation of several objects in the environment may have a direct influence on the behavior of the robot. Typical objects whose states are relevant for the robot's actions are doors, tables, wastebaskets, and so forth. To accurately deal with such dynamic objects, their current states have to be visualized. The RTL-system therefore includes a probabilistic technique to estimate continuously the states of these objects based on the data obtained with the robot's sensors. It determines the current object states (position and/or orientation) and transfers them to all sites connected over the Internet, which instantly update the visualization accordingly.

This chapter presents the architecture of the RTL-system. It describes the predictive simulation technique as well as the state estimation method. After discussing related work in the following section, the key ideas of the visualization system are described in section 14.2. Section 14.3 presents several experiments illustrating the advantages and the robustness of this approach.

14.1.1 Related Work

A variety of online teleoperation interfaces for robots has been developed over the last few years. Three of the earlier systems are the Mercury Project, the "Telerobot on the Web," and the Telegarden [8, 9, 23]. These systems allow people to perform simple tasks with a robot arm via the Web. Since these arms operate in prepared workspaces without any unforeseen obstacles, all movement commands issued by a web user can be carried out in a deterministic manner. Additionally, it suffices to provide still images from a camera mounted on the robot arm after a requested movement task has been completed. The system described here, in contrast, is designed to visualize the actions of an autonomous mobile robot operating in a dynamic environment. Accordingly, the execution of the same task depends on the current situation in the robot's environment.

Xavier [22] is a web-operated autonomous mobile robot. It can be instructed by web users to move to certain offices and to tell a joke after arrival. The web interface relies on client-pull and server-push techniques to provide visual feedback of the robot's movements, including images taken by the robot as well as a map indicating the robot's current position. Xavier's interface, however, does not include any techniques for bridging transmission gaps or to reflect changes of the environment.

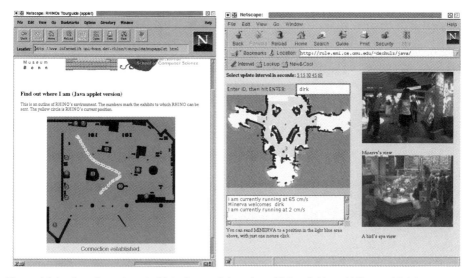

Figure 14.1:   Java-based control interfaces of the robots Rhino (left) and Minerva (right).

The autonomous mobile robots Rhino and Minerva, which were deployed as interactive museum tour-guide robots in the Deutsches Museum Bonn in 1997 and in the Smithsonian Museum of American History in 1998, could also be operated over the Web [2, 24]. In addition to image streams, recorded with on-board and off-board cameras, their interfaces offered Java applets for instant updates of information. For example, the interfaces displayed information about the current actions of the robot and included applets providing smooth 2-D animations of the robot's trajectory in a map of the environment [22] (see figure 14.1). Although both robots used the communication interface of the RTL-system, the interfaces did not include any means to compensate transmission gaps. The map employed by the visualization was completely static, and dynamic aspects were not handled at all.

3-D graphics visualizations for Internet-based robot control have already been suggested by Hirukawa et al. [10]. Their interface allows web users to carry out manipulation tasks with a mobile robot by controlling a 3-D graphics simulation of the robot contained in the web browser. In contrast to the RTL-system described here, these interfaces do not provide online visualizations of the robot's actions. Instead, they follow the tele-programming approach. Tasks are first tested off-line in the 3-D simulation environment and are afterward transmitted via the Internet for execution by the real robot.

Simulation-based delay compensation techniques and virtual environment interfaces are frequently used for space and underwater robotics [3, 26, 21, 1]. These methods are designed to deal with large transmission delays but, in contrast to Internet based systems, rely on communication links with a guaranteed bandwidth and known transmission gaps. A visualization method, which is often used in telemanipulation systems, is predictive displays [13]. Methods of this type combine low-bandwidth video transmissions with graphics visualizations.

The simulation scheme used here is conceptually similar to the one used during the ROTEX experiment on board the space-shuttle COLUMBIA in 1993 [11]. During this experiment a 6 DOF robot arm was controlled from ground employing telesensor programming. A human operator planned a sequence of elemental operations that afterward was carried out by the robot. Elemental operations are short-term reactive control tasks, like grasping operations, which the robot can carry out autonomously using its sensors. Truly autonomous robots, however, include methods to plan even complex actions and to react to unforeseen circumstances. For example, most environments are not completely accessible. A mobile robot therefore must have the ability to autonomously plan its path to a given target point. In addition, it must be able to replan dynamically whenever unexpected obstacles block its path.

The predictive simulation system described in this chapter is designed to provide accurate and detailed visualizations of the actions carried out by a mobile robot over the Internet. The main advantage of this system is that it includes techniques to deal with unreliable connections without a guaranteed bandwidth such as the Internet. In the case of transmission gaps, this RTL-system uses a robot simulator to simulate the robot's actions. It also uses path planning components to accurately forecast the robot's actions even if transmission gaps of more than 10 seconds occur.

In general, it is not sufficient to regard the environment as static and to display the robot within a static model of the environment. In practice, the environment of a mobile robot contains several nonstatic objects that influence the behavior of the robot; among them are doors, chairs, and tables, which often change their position and can prevent the robot from taking a previously planned trajectory. To acquire and maintain a model of the environment is a major research area in mobile robotics. The most frequently used types of maps are metric and topological maps. Topological models, as used in [14, 17], describe the environment at a coarse resolution. Because of the lack of necessary details, these types of maps are only of limited use for visualizations. Metric maps, on the other hand, describe the environment at a finer level of detail. A popular approach is discrete occupancy grids proposed in [5, 16]. Each cell of such a grid contains the probability that the corresponding space in the environment is occupied. The major advantage of occupancy maps lies in the existence of techniques for their acquisition and maintenance based on sensor information. Most of the occupancy grid techniques, however, are intended for static environments only. Since all cells are considered independently, they cannot appropriately represent dynamic objects such as doors, tables, or chairs. The RTL-system uses a 3-D model of the environment containing all relevant objects of the environment. A probabilistic approach is applied to estimate the state of dynamic objects. Changes of these states are also updated in the 3-D model so that the visualization can quickly be adapted to the current state of the environment.

14.2   The Visualization System

To display the robot and its actions in the environment a 3-D visualization system is used. This visualization system requires a 3-D model of the robot's environment represented by a hierarchical scene graph in which objects are described in the boundary representation. Figure 14.2 shows the 3-D model of the department in which the experiments described below were carried out. Given a corresponding description of the robot, it can provide a visualization of the robot at arbitrary locations in the environment.

The RTL-system has a client-server architecture. Initially, each client receives the 3-D model. Afterward, the server sends updates of the current state of the robot in regular intervals to all its clients. Unfortunately, the bandwidth of the Internet is limited and the delivery of packets is unreliable so that larger transmission gaps can occur especially on overseas connections. To cope with these limitations, the RTL-system provides a predictive simulation to predict the robot's actions even when larger transmission gaps of several seconds occur.

As pointed out above, the environment of the robot is not static. Generally, there are several dynamic objects in the real world that influence the robot's behavior. To correctly reflect the state of these objects, the remote clients also receive the states of these objects from the server to update the world model

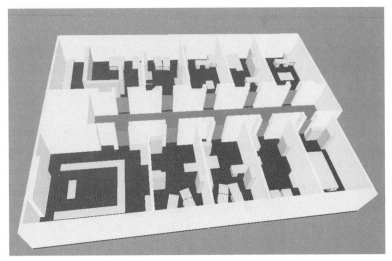

Figure 14.2: Three-dimensional boundary representation of the robot's environment.

according to changes. To provide accurate states of objects, the server is connected to a state estimation component, which continuously determines the state of objects using the robot's sensors.

14.2.1 State Estimation of Dynamic Objects

To estimate the state of dynamic objects, a mobile robot has to use its sensors. The information provided by the sensors, however, is inherently noisy so that state estimates relying on this information are always to some degree uncertain. Estimating the state of objects based on sensor information provided by a mobile robot furthermore depends on the robot's position within the environment. This position also has to be determined by a state estimation process and therefore is also uncertain. To correctly deal with these uncertainties, probabilistic state estimation techniques are used here.

The correct probabilistic approach to continuously maintain a dynamic world model is to maintain the joint distribution over all states of the dynamic objects within the environment. Unfortunately, this approach is not tractable since the size of the joint state space grows exponentially in the number of objects. Therefore, a marginal distribution was considered and the states of the dynamic objects are estimated independently.

Since the measurements o of the robot are inherently uncertain, one is interested in the most likely state $\tilde{s}$ of each object, that is the state which maximizes the probability $p(s|o)$

$$\tilde{s} = \underset{s}{\mathrm{argmax}}\ p(s|o) \tag{14.1}$$

Here $p(s|o)$ is the probability that the currently considered object is in state s given the observation o.

To compute $p(s|o)$ one has to take into account that the position l of the robot from which the observation o is obtained is generally uncertain. Here it is assumed that the robot maintains a belief $p(l)$ over all possible locations. To take the current belief of the robot into account, one integrates over all possible locations of the robot:

$$p(s|o) = \int_l p(s|o, l) \cdot p(l) dl \tag{14.2}$$

Given Bayes rule using l as background knowledge one obtains

$$p(s|o, l) = \frac{p(o|s, l) \cdot p(s|l)}{p(o|l)} \tag{14.3}$$

where the denominator $p(o|l)$ can be rewritten as

$$p(o|l) = \int_s p(o|s, l) \cdot p(s) ds \tag{14.4}$$

Inserting equation (14.3) into equation (14.2) leads to

$$p(s|o) = \int_l \frac{p(o|s, l) \cdot p(s|l)}{p(o|l)} \cdot p(l) dl \tag{14.5}$$

Under the strong assumption that the state s of the object and the position l of the robot are independent, one simplifies this to

$$p(s|o) = \int_l \frac{p(o|s, l) \cdot p(s) \cdot p(l)}{p(o|l)} dl \tag{14.6}$$

Thus, all one needs to know are the quantities $p(s)$ which is the prior distribution of the state s of the currently considered object, the current belief $p(l)$ of the position of the robot as well as the term $p(o|s, l)$.

To estimate $p(l)$, that is, the probability that l is the robot's current position, the RTL-system applies Monte Carlo localization [4, 6] which is an efficient variant of Markov localization [7]. In several robot systems applied in real-world scenarios, Markov localization has been proven to be robust and to provide accurate position estimates. The key idea of Markov localization is to maintain a probability distribution over all possible positions of the robot. Monte Carlo localization is a variant, which uses a set of samples to represent this density. The RTL-system uses this sample set as input for the state estimation process. Since the samples provided by Monte Carlo localization are randomly drawn from the probability density about the robot's current position, equation (14.6) can be reduced to

$$p(s|o) = \frac{1}{|\Lambda|} \sum_{\lambda \in \Lambda} \frac{p(o|s, \lambda) \cdot p(s)}{\int_s p(o|s, \lambda) \cdot p(s)ds} \qquad (14.7)$$

where Λ is the set of samples representing the robot's current belief about its position.

The term $p(o|s, l)$ belongs to the crucial aspects of the state estimation process since it describes the probability of making a certain observation o given the state s of the object and the location l of the robot. The usual way to compute this probability is a template-matching process that compares the current measurement to the expected measurement given the position of the robot and the state of the object. The current version of the RTL-system uses a laser range finder for this purpose. A scan of a laser range sensor generally consists of a set of beams representing the distance to the next obstacles at a fine angular resolution (see, for example, the left image of figure 14.4). Under the assumption that the individual beams are independent given the position of the robot, one can compute $p(o|s, l)$ in the following way:

$$p(o|s, l) = \prod_{i = 1, ..., n} p(d_i|\hat{d}_i(s, l)) \qquad (14.8)$$

Here n is the number of beams, d_i is the length of beam i, and $\hat{d}_i(s, l)$ is the expected distance for beam i given the state s of the object and the location l of the robot. In the current implementation, a Gaussian distribution centered around the expected distance $\hat{d}_i(s, l)$ is used to determine the probability $p(d_i|\hat{d}_i(s, l))$.

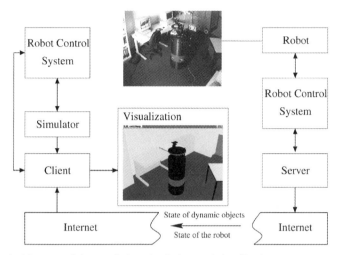

Figure 14.3: Architecture of the predictive simulation and visualization system.

14.2.2 Predictive Simulation

As pointed out, one of the major drawbacks of the Internet in the context of teleoperated mobile robots is the missing guarantee of a certain transmission rate and the loss of packets. Thus, it is generally impossible to obtain updates of the current position of the robot with a guaranteed frequency. To cope with this problem, visualization techniques have to extrapolate the actions of the robot between two subsequent packets to provide smooth animations of the robot's actions. In this context, a common technique is to extrapolate the robot's actions using the current velocities, which directly correspond to performing dead reckoning based on the latest robot position and velocities received. The RTL-system, in contrast, uses a predictive simulation technique, that also respects the real robot's path-planning component to achieve better long- term predictions. As a result, the RTL-system provides accurate predictions for more than 10 seconds.

Figure 14.3 shows the client-server architecture of the predictive simulation system. The server, which is located at the robot's site, sends synchronization messages via the Internet to its clients. Each synchronization message contains the current state of the robot consisting of the current position, velocities, and target point of the robot. It also includes the current state of the dynamic objects as well as the current time.

Instead of the standard TCP protocol utilized by most applications, the RTL-system uses the unreliable UDP protocol to transfer the state information from the server to its clients. Since the Internet is subject to packet loss, the use of a reliable protocol like TCP can increase transmission delays, as the retransmission of lost messages prevents the arrival of new information. In several experiments carried out with typical long-distance connections, packet loss rates of up to 30 percent were frequently observed, whereas the transmission time of a single packet rarely exceeded 0.3 seconds. To deal with longer transmissions, each client warns the user if a packet needed more than 0.4 seconds the delivery. The synchronization of the clocks of the clients and the server is obtained by the *network time protocol* [15].

After receiving a state message, the client updates the position of the robot and the state of the dynamic objects. Afterward it starts a simulation to predict the robot's actions until the next packet is received. The current implementation of the system [19] includes a simulator to simulate the robot's actions and its proximity sensors. It is also connected to the path planning components [25] used by the real robot.

The simulation component provides methods for carrying out ray casting within the world model, which builds the basis for the proximity sensor simulation required to simulate the behavior of the path planning components. Since realistic scene descriptions generally contain a large number of objects, a huge number of line-polygon intersections have to be calculated to simulate the proximity sensors. To achieve this in real-time, the simulation system applies a spatial indexing technique based on the rectangular tiling of the scene [18].

14.3   Experimental Results

The state-estimation technique and the predictive-simulation method have been implemented and extensively tested and evaluated using the mobile robot Rhino, an RWI B21 robot. Rhino is equipped with two laser range finders covering almost 360 degrees of the robot's surrounding. These sensors are used for the evaluation of the state-estimation technique described above. The visualization system is based on OpenGL so that the actions of the robot can be visualized with at least 20 Hz frame rate using 3-D hardware acceleration on a standard low-cost PC.

Figure 14.4: Rhino detecting the state of two doors with its laser range-finder (left image). The estimated states of the doors are depicted in black. The resulting 3-D visualization is shown in the right image.

14.3.1 State Estimation for Dynamic Objects

To evaluate the state-estimation technique and its advantages for a robust visualization of the robot's activities, several experiments were performed in an office environment. In these experiments, the task of the robot was to estimate the state of the doors while the robot was traveling along the 26m-long corridor of the department building.

To efficiently carry out the state-estimation process, the current implementation uses a discrete set of possible states. For doors, typically, 10 states ranging from 0 to 90 degrees were used. To obtain more accurate estimates of the state of objects, a small set of different measurements were integrated. Figure 14.4 shows a typical situation in which the robot measures two doors. The resulting states of the doors are depicted in black. Figure 14.4 also shows the corresponding 3-D visualization.

Figure 14.5 illustrates the trajectory of the robot as well as the states of the doors which were kept unchanged during the experiment. Table 14.1 shows the results of the state estimation process for each door. As can be seen from these data, the RTL-system provides highly accurate results. In most of the cases the standard deviation is below 5 degrees.

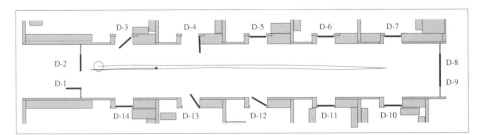

Figure 14.5: *Trajectory of the robot and state of the doors in the corridor of the Department of Computer Science, Bonn. Figure 14.7 shows the resulting 3-D visualization.*

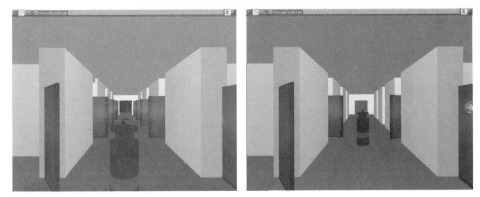

Figure 14.7: *Visualization of the corridor at the beginning (left image) and at the end of the trajectory (right image).*

Table 14.1: *Estimates of the state of the doors depicted in figure 14.5.*

Door	1	2	3	4	5	6	7	8	9	10	11	12	13	14
m	180	270	220	264	182	180	182	270	90	176	178	155	124	173
s	0.0	0.0	5.3	6.4	5.0	0.0	4.3	0.0	0.0	4.6	3.9	5.0	7.1	6.5
N	21	25	43	23	17	17	33	27	26	37	13	15	19	41
Exact	180	270	218	270	180	180	180	270	90	180	180	150	122	180

14.3.2 Predictive Simulation

To demonstrate the prediction accuracy of the predictive-simulation system, several navigation experiments were performed with the mobile robot Rhino. The goal of these experiments was to demonstrate the advantages of the predictive-

Figure 14.8: Trajectory of the robot (left image), trajectory computed by dead reckoning (center image), and trajectory computed by the predictive simulation technique (right image).

simulation system over a visualization technique based on dead reckoning given the latest state information received from the server.

In the first experiment, Rhino started in the corridor of the department and moved into five different offices in the order indicated by the numbers in the left image of figure 14.8. The length of the trajectory is 80.8 m and the robot needed 320 seconds to complete this task.

To evaluate the capabilities of the predictive simulation system, the packet transmission rate was manually decreased to approximately ten packets per minute. A constant transmission interval was used, so that the time gap between subsequent packets was approximately 6 seconds. Since packets sometimes get lost, the effective time-gap between subsequent packets in several cases was $6n$ seconds. The center plot of figure 14.8 shows the trajectory obtained if the robot's movements are extrapolated using dead reckoning. Obviously the resulting trajectory differs significantly from the true trajectory of the robot. Furthermore, it contains several long gaps of more than 1m illustrating the limited quality of the standard dead reckoning approach. The predictive simulation system, however, provides a trajectory that is much closer to the robot's actual trajectory (see right image of figure 14.8).

To demonstrate the improvements in the visualization, figure 14.9 illustrates a small fraction of the trajectory of the robot after it left room It shows the corresponding trajectory of the robot (left image), the trajectory obtained by extrapolation (center image), and the trajectory computed by the predictive visualization system (right image). In this particular case one packet was lost so that the system had to predict the robot's trajectory for a period of 12.5 seconds. Since the dead reckoning does not change the velocities of the robot during a

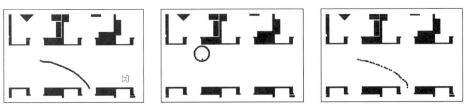

Figure 14.9: Trajectory of the robot taken after leaving room 1 (left), circular trajectory obtained by extrapolation (center), and trajectory extrapolated by the predictive simulation (right).

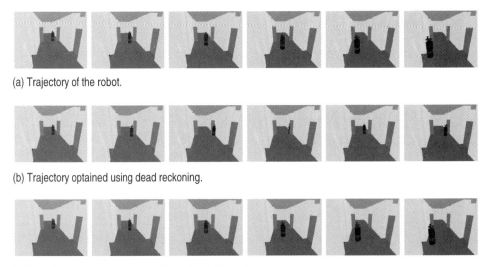

(a) Trajectory of the robot.

(b) Trajectory optained using dead reckoning.

(c) Trajectory computed by the predictive simulation system.

Figure 14.10: A sequence of computed images showing Rhino moving through the corridor of the department. The time difference between consecutive images is approximately 2 seconds. The camera position is illustrated in the left image of figure 14.9.

transmission gap, the robot moves on a circular trajectory. In contrast, the predictive-simulation system uses the path-planning components to forecast the actions of the robot. Therefore, the resulting trajectory is much closer to that of the real robot.

The corresponding sequences of images of the 3-D visualization are shown in figure 14.10. The time delay between consecutive frames is 2 seconds. The first row shows the images obtained using the correct position of the robot. The second row contains the images computed by the dead-reckoning extrapolation, and the

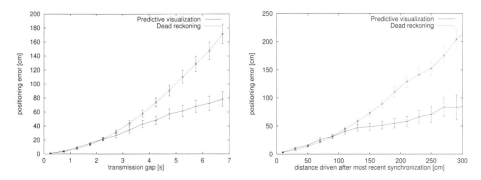

Figure 14.11: Average displacement depending on the length of the transmission gap (left image) and on the distance traveled after the last synchronization (right image).

third row includes the images obtained using the predictive-visualization system. Obviously, the predictive-visualization system significantly improves the quality of the visualization.

To obtain a quantitative assessment of the prediction quality, the whole experiment was repeated several times using the data shown in the left image of figure 14.8. In each individual run the packets which were not transmitted were randomly selected. The left plot in figure 14.11 shows the average positioning error — the distance between the predicted position and the true position of the robot — as a function of transmission gap. Here, the predictive-visualization system is able to significantly reduce the average positioning error compared to dead reckoning after transmission gaps of at least 2.5 seconds. Since these differences depend on the speed of the robot, which was $34.5ms^{-1}$ on average during this experiment, the average displacement was also computed depending on the distance traveled after the latest synchronization packet. The resulting plot is shown in the right half of figure 14.11. Obviously, after at least 1m of travel, the predictive visualization system provides significantly better position estimates. In both figures, the error bars indicate the 95 percent confidence interval of the average mean.

14.4   Conclusions

This chapter presents an approach for the robust 3-D visualization of the actions of a mobile robot over the Internet. Instead of image streams, the technique used here includes a 3-D model of the environment and visualizes the actions of the robot and the current state of the environment by 3-D rendering. Accordingly, only the current state of the robot and of dynamic objects within the environment have to be transferred over the Internet which results in a serious data reduction. This approach applies a probabilistic state estimation technique to continuously determine the current states of dynamic objects. It also includes a predictive simulation component to bridge transmission gaps of several seconds, which frequently occur when the Internet is used as the communication medium.

This system has been implemented and tested extensively in a typical office environment. The results show that the predictive simulation provides significantly more reliable 3-D visualizations than the standard dead-reckoning approach, which forecasts the actions of the robot based on the latest state information. Furthermore, it was demonstrated that the state estimation procedure can robustly and efficiently estimate the state of dynamic objects in the environment of the robot. As a result, accurate visualizations of the robot and its environment were obtained even if larger transmission gaps of 10 seconds occur.

There are several aspects for future research. For example, using 3-D visualizations has the advantage that a user can watch the robot from arbitrary perspectives by selecting different viewpoints for the virtual camera. Choosing appropriate view-points, however, is not at all an easy task, especially for nonexpert users. The RTL-system therefore requires techniques for selecting viewpoints automatically and appropriate user interfaces. One approach toward this direction is described in [12].

Since the robots operate in populated environments, the authors would also like to include techniques for detecting and tracking people in the robot's surrounding. In the first experiments it was found that the template-matching approach presented here is not adequate for this task, especially if people are walking through the robot's environment. Moreover, since people often show up in groups, they cannot be tracked independently. In this context, techniques known from multiple hypothesis tracking appear to be promising approaches.

Acknowledgments

This project has been supported by the Northrhine-Westphalia Ministry for Schools, Education, Science and Research (MSWWF-NRW, Germany), under contract number IVA3-10703596 and by the IST Programme of Commission of the

European Communities under contract number IST-1999-12643. The authors would like to thank Dieter Fox, Michael Beetz, and Sebastian Thrun as well as the members of the research lab for autonomous mobile systems at the University of Bonn for fruitful discussions.

References

[1]   B. Bon and S. Homayoun. Real-time model-based obstacle detection for the NASA ranger telerobot. In *Proc. of the IEEE International Conference on Robotics & Automation (ICRA)*, April 1997.

[2]   W. Burgard, A. B. Cremers, D. Fox, D. Hähnel, G. Lakemeyer, D. Schulz, W. Steiner, and S. Thrun. Experiences with an interactive museum tour-guide robot. *Artificial Intelligence*, 114 (1 – 2), 1999.

[3]   L. Conway, R. A. Volz, and M. W. Walker. Teleautonomous systems. Projecting and coordinating intelligent action at a distance. In *IEEE Transactions on Robotics and Automation,* vol. 6, April 1990.

[4]   F. Dellaert, D. Fox, W. Burgard, and S. Thrun. Monte Carlo localization for mobile robots. In *Proc. of the IEEE International Conference on Robotics & Automation (ICRA)*, 1999.

[5]   A. Elfes. Sonar-based real-world mapping and navigation. *IEEE Transactions on Robotics and Automation 3* (3): 249 – 265, 1987.

[6]   D. Fox, W. Burgard, F. Dellaert, and S. Thrun. Monte Carlo localization: Efficient position estimation for mobile robots. In *Proc. of the National Conference on Artificial Intelligence (AAAI)*, 1999.

[7]   D. Fox, W. Burgard, and S. Thrun. Markov localization for mobile robots in dynamic environments. *Journal of Artificial Intelligence Research 11,* 1999.

[8]   K. Goldberg, M. Mascha, S. Gentner, N. Rothenberg, C. Sutter, and J. Wiegley. Desktop tele-operation via the world wide web. In *Proc. of the IEEE International Conference on Robotics & Automation (ICRA)*, 1995.

[9]   K. Goldberg, J. Santarromana, G. Bekey, S. Gentner, R. Morris, J. Wiegley, and E. Berger. The telegarden. In *Proc. of ACM SIGGRAPH,* 1995.

[10] H. Hirukawa, T. Matsui, and H. Onda. Prototypes of teleoperation systems via a standard protocol with a standard human interface. In *Proc. of the IEEE International Conference on Robotics & Automation (ICRA),* 1997.

[11] G. Hirzinger, B. Brunner, J. Dietrich, and J. Heindl. Sensor-based space robotics-ROTEX and its telerobotic features. In *IEEE Transactions on Robotics and Automation*, vol. 9, October 1993.

[12] A. Hopp, D. Schulz, W. Burgard, A. B. Cremers, and D. Fellner. Virtual reality visualization of distributed tele-experiments. In *Proc. of the 24th Annual Conference of the IEEE Industrial Electronics Society (IECON)*, 1998.

[13] W. S. Kim and A. K. Bejczy. Demonstration of a high-fidelity predictive/ preview display technique for telerobotic servicing in space. In *IEEE Transactions on Robotics and Automation,* vol. 9, October 1993.

[14] D. Kortenkamp and T. Weymouth. Topological mapping for mobile robots using a combination of sonar and vision sensing. In *Proc. of the Twelfth National Conference on Artificial Intelligence,* p. 979 – 984, 1994.

[15] D. L. Mills. Network time protocol (version 3) specification, implementation and analysis. In *RFC 1305,* Internet Engineering Task Force (IETF), March 1992.

[16] H.P. Moravec. Sensor fusion in certainty grids for mobile robots. *AI Magazine,* summer 1988.

[17] D. Pierce and B. Kuipers. Learning to explore and build maps. In *Proc. of the Twelfth National Conference on Artificial Intelligence.* AAAI Press, 1994.

[18] H. Samet. *Applications of Spatial Data Structures.* Addison-Wesley Publishing Company, 1990.

[19] D. Schulz, W. Burgard, and A. B. Cremers. Robust visualization of navigation experiments with mobile robots over the internet. In *Proc. of the IEEE/RSJ International Conference on Intelligent Robots and Systems (IROS),* 1999.

[20] D. Schulz, W. Burgard, D. Fox, S. Thrun, and A. B. Cremers. web interfaces for mobile robots in public places. *IEEE Robotics & Automation Magazine,* volume 7 (1), March 2000.

[21] T. B. Sheridan. Space teleoperation through time delay: Review and prognosis. In *IEEE Transactions on Robotics and Automation,* vol. 9, October 1993.

[22] R. Simmons, R. Goodwin, K. Haigh, S. Koenig, and J. O'Sullivan. A layered architecture for office delivery robots. In *Proc. of the First International Conference on Autonomous Agents,* Marina del Rey, CA, February 1997.

[23] K. Taylor and J. Trevelyan. A telerobot on the World Wide Web. In *Proc. of the 1995 National Conference of the Australian Robot Association,* 1995.

[24] S. Thrun, M. Bennewitz, W. Burgard, A. B. Cremers, F. Dellaert, D. Fox, D. Hähnel, C. Rosenberg, N. Roy, J. Schulte, and D. Schulz. MINERVA: A second generation mobile tour-guide robot. In *Proc. of the IEEE International Conference on Robotics & Automation (ICRA),* 1999.

[25] S. Thrun, A. Bücken, W. Burgard, D. Fox, T. Fröhlinghaus, D. Hennig, T. Hofmann, M. Krell, and T. Schimdt. Map learning and high-speed navigation in RHINO. In D. Kortenkamp, R.P. Bonasso, and R. Murphy, ed. *AI-based Mobile Robots: Case studies of successful robot systems.* Cambridge: MIT Press, 1997.

[26] Y. Tsumaki and M. Uchiyama. A model-based space teleoperation system with robustness against modeling errors. In *Proc. of the IEEE International Conference on Robotics & Automation (ICRA),* 1997.

Part IV

Other Novel Applications

15 Computer Networked Robotics

Hideki Hashimoto and Yasuharu Kunii

15.1 Introduction

Recently, multimedia applications and the Internet have attracted the attention of the world. Many kinds of information run through this computer network, but information on the network cannot have a physical influence on the real environment. Robots, however, can expand the ability of multimedia to have a physical impact [1, 2, 3]. Robots can behave as physical media or agents that have an effect on the physical environment through the network. With the spreading availability of high bandwidth networks, B-ISDN connections will be broadly available early in this century. The authors are currently building a system called Robotic Network System (RNS) in which robots can be physical agents as one of the media. In such a system various types of robots are connected to this network in different places such as homes, schools, hospitals, and fire and police stations. And robots are going to come into our daily life.

This chapter introduces two experimental systems and experimental results of RNS. One is a mobile-robot: Multi-Sensored Autonomous Rover (MUSEAR), which behaves as a physical agent of an operator in a remote site. The operator can drive MUSEAR as an agent, watching a real-time image from its on-site camera in the remote site. The other is the HandShake Device (HSD), which is a haptic interface using teleoperation and force-feedback techniques to display physical information through the network [4]. This system is a testbed of teleoperation, and it is defined as a physical media using force information.

15.2 Robot as Physical Agent

A remote environment can be accessed through the computer network by using a robot. It will be used to expand the limitations of human ability, for example time and physical constraints. This section, the first step of the research, shows teledriving experimental results of the mobile robot MUSEAR.

15.2.1 MUSEAR: Physical Agent

The mobile robot, MUSEAR, is shown in figure 15.1. MUSEAR is defined as a

Figure 15.1: Multi-Sensored Autonomous Rover: MUSEAR.

home robot in RNS, and it is based on an electric wheel chair. It can have autonomous navigation with a note-type PC and some sensors, for example, ultrasonic sensors. This system communicates to other computer systems through the Internet by a wireless LAN, and it can take information from them, for example, map information.

15.2.2 Teledriving System and Experimental Results

In the experiment of teledriving, MUSEAR is connected to the host through the Internet between the Institute of Industrial Science (Roppongi in Tokyo) and Waseda University, a distance that takes 40 minutes by train from Roppongi. Two physical agents operated at that time. One was the system MUSEAR. The other was the Mini Planetary Rover (MPR) of Meiji University. Each system could recognize the other through a camera on each system. The experimental system structure of MUSEAR is shown in figure 15.2. An operator uses an analog joystick to control a velocity vector while watching a color video image transmitted by teleconference system. Resource Reservation Protocol (RSVP) was used to keep data speed (64Kbps) on some network routers. Figure 15.3 (a) and (b) shows a photograph of the experiment and its control screen. An interaction with MPR was located in the remote environment. Operators of each system can recognize the other, as shown in figure 15.3 (b) from MUSEAR and (c) from MPR.

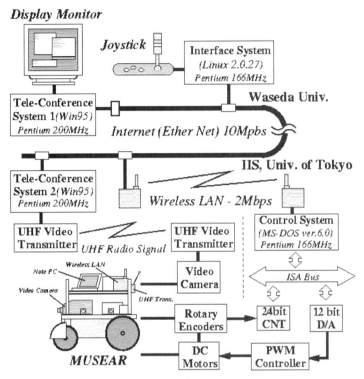

Figure 15.2: Experimental system for teledriving.

In this experiment, those systems could bee smoothly controlled in the morning but not in the afternoon because of light communication load. The difficulty in operation is mainly caused by a communication delay of image data. The delay of image data is more important than of command data because a communication delay is not as large in this case, so there was a need to improve GUI with some assistance to the host. The variable time-delay on the network is also one of the reasons for the difficulty of an operation, and it always exists on the network. The compensation of this variable delay will be one for future work.

15.3   Robot as Physical Media

It is possible to transmit physical information such as a force and torque information from a remote site by using the network and robot technology. For example, a haptic interface as a master system of teleoperation is a media that

(a) Operator for MUSEAR

(b) Image from MUSEAR

(c) Image from MPR of Meiji University

Figure 15.3: Experiment of telecontrol between IIS and Waseda University.

displays a remote environment by force information. The haptic interface here is called HandShake Device (HSD), described in this section, and it has shown this possibility.

15.3.1 HandShake Device: Physical Media

HSD is an interface device that transfers the motion of the handshake partner to the user. Therefore HSD represents the hand of the other user as shown in figures 15.4 and 15.5.

Figure 15.6 shows the structure of HSD, which is composed of two arms: the master

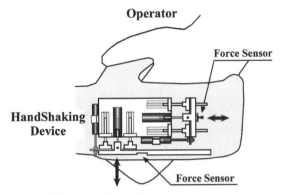

Figure 15.4: Handshake via HSD (HandShake Device).

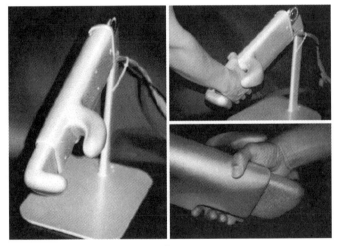

Figure 15.5: Photograph of Hand Shake with HSD.

arm and the slave arm. The master arm is grasped by the user while the slave arm represents the handshaking action done by the other user (handshake partner). Each arm is connected to the shaft of the linear motion motor [5]. The linear guiders support linear motion of the arms. Positions of the master and slave arms are measured by linear-motion potentiometers. Force information is acquired via a strain gauge bridge connected to each arm (figure 15.7).

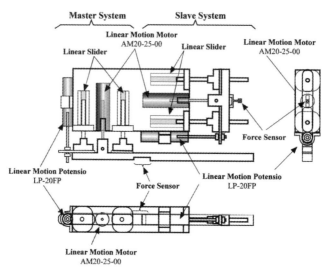

Figure 15.6: Schematic of HSD.

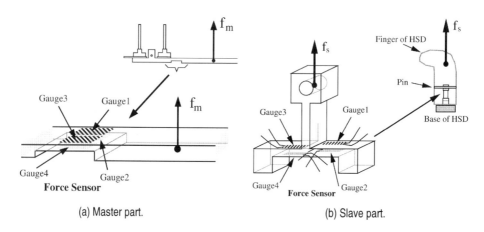

(a) Master part.   (b) Slave part.

Figure 15.7: Schematic of Force Sensor on Master and Slave parts on HSD.

15.3.2   System Structure of HSD

The telehandshaking system contains two HSDs (one located at site A and operated by operator A; the other located at site B and operated by operator B) and it is composed of two single degree-of-freedom (DOF) linear motion master-slave

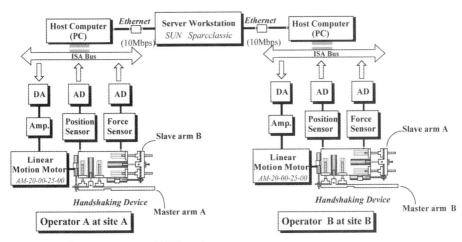

Figure 15.8: System structure of HSD on Internet.

systems: (1) the master arm A and the slave arm A operated by the operator A, and (2) the master arm B and the slave arm B operated by the operator B. The structure of the telehandshaking system is illustrated in figure 15.8.

At each site, position and force information from position and force sensors are sent through an AD board to a host computer (PC Pentium 100 MHz). The controller is realized in the host computer. The controller generates control signals that are sent through a DA board and an amplifier to the motors to control the motion of the master and slave arms.

Data between two sites are communicated through the Internet. The data from one host computer are transmitted to the other host via a server workstation and vice versa. The server workstation can be located at some place in the network to set node for passing data when it is sent through the Internet by the most standard network protocols TCP/IP.

15.3.3 Control Algorithm for HSD

Virtual Impedance

Based on the concept of the impedance control, a technique of compliant motion for a teleoperator by using a virtual impedance (VI) (or virtual internal model or virtual model) has been used [6, 7, 8].

It is assumed that an end-point (or end-effector) of a manipulator arm in a

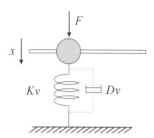

Figure 15.9: *Structure of virtual impedance for*
1 DOF linear motion manipulator.

virtual compliance frame is controlled in a position depending on the virtual impedance of the frame. VI is composed of a virtual spring (or compliance), a virtual damper, and a virtual mass. These parameters may be programmed and/or be different in direction, or change with time to produce good maneuverability.

In the case of a 1 DOF linear motion manipulator, it is assumed that a virtual mass, M_v, is attached to a manipulator arm. This virtual mass is supported by a virtual damper, D_v, and a virtual spring, K_v. The structure of the VI is shown in figure 15.9.

When a force is applied to the virtual mass, the position of the virtual mass is modified according to the dynamics of VI. The dynamic equation of VI is as follows.

$$F(t) = M_v \ddot{x}(t) + D_v \dot{x}(t) + K_v x(t) \tag{15.1}$$

where $F(t)$ is the force detected at the virtual mass and $x(t)$ is the displacement of the virtual mass.

Compensation of Time-delay

Communication time-delay comes out on a teleoperation system. Furthermore the Internet has variable time-delay because of network load by its own anonymous users.

Variable time-delays on the Internet (Ether) and the B-ISDN (ATM network) between Tokyo and Nagoya are shown in figure 15.10. Some large time-delays can be recognized in figure 15.10. The time-delay of a ATM network is better than Ether-net, however, it still has a 100[ms] delay. Variable time-delays might be an important problem for teleoperations on a computer network. Here, variable time-delay will be a future work to study and it is assumed that time-delay is constant in this research.

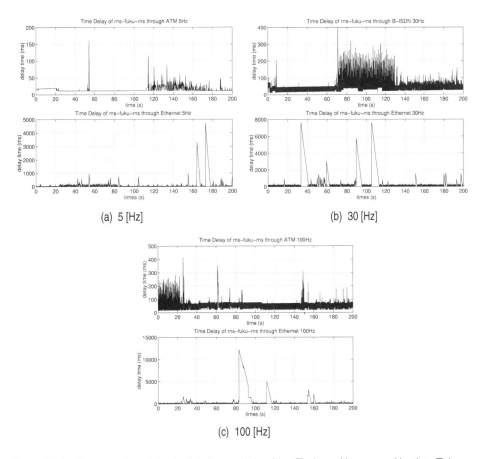

Figure 15.10: One-way time delay in data transmission from Toyko to Nagoya and back to Tokyo.

A new control approach, virtual impedance with position error correction (VIPEC), is proposed [9]. This new control approach is an extension of the control approach by Otsuka et al. [8]. The control diagram of a teleoperator with time-delay utilizing the VIPEC is illustrated in fig. 15.11. The PEC part (the shaded part) is added to the original approach to improve the response of the teleoperator by reducing the position error between the master and slave positions x_m, x_s.

Modeling of HSD

Figure 15.11 shows the control-system diagram of the master-slave system inside the telehandshaking system with VIPEC approach. The armature inductance, L_a

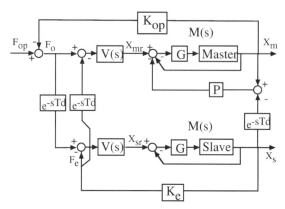

Figure 15.11: Control system diagram with VIPEC for HSD.

of each motor is neglected. The two linear motion motors represented by the master block and the slave block can be modeled as simple DC-motors with the transfer function

$$C(s) = \frac{X(s)}{E_a(s)} = \frac{K_m}{s(T_m s + 1)} K_p \tag{15.2}$$

where X is the displacement of the motor shaft, E_a is the applied armature voltage, K_m is the motor gain constant, T_m is the motor time constant, and K_p is a constant to convert the angular displacement into displacement. K_m and T_m are obtained from

$$K_m = \frac{K}{R_a + KK_b} \quad ; \quad T_m = \frac{R_a J}{R_a b + KK_b} \tag{15.3}$$

where K is the motor-torque constant, R_a is the armature resistance, K_b is the back emf constant, J is the moment of inertia of the motor, and b is the viscous-friction of the motor which is very small and can be neglected.

The servocontroller is a proportional controller whose gain is G. The closed-loop transfer function of the manipulator arm and the controller is

$$M(s) = \frac{C(s)G}{1 + C(s)G} = \frac{\varphi}{\alpha s^2 + \beta s + \gamma} \tag{15.4}$$

where $\alpha = 1$, $\beta = 1/T_m$, and $\gamma = \varphi = (GK_m K_p)/T_m$.

Gain P represents PEC gain. For simplicity, the environment is modeled by a spring with stiffness constant K_e. Therefore the dynamics of the spring interacting with the slave arm is modeled as follows.

$$F_e(t) = K_e x_s(t) \tag{15.5}$$

The dynamics of the operator including the dynamics interaction between the operator and the master arm, is approximated by

$$F_{op}(t) - F_o(t) = M_{op} \ddot{x}_m(t) + D_{op} \dot{x}_m(t) + K_{op} x_m(t) \tag{15.6}$$

where F_{op} is the force generated by the operator's muscles. M_{op}, B_{op}, and K_{op} denote mass, damper, and stiffness of the operator and the master arm respectively.

Table 15.1: Constants of the system model.

Armature resistance, R_a	10.9 [W]
Armature inductance, L_a	neglected
Viscous friction, b	neglected
Moment of inertia, J	$2.08\ 10^{-7}$ [kg m^2]
Back emf constant, K_b	0.034 [V/(rad/s)]
Motor-torque constant, K	0.06 [Nm/A]
Motor time constant, T_m	0.0011 [s]
Controller gain, G	10000
Motor gain constant, K_m	29.41 [(rad/s)/V]
Environment stiffness, K_e	1200 [N/m]
Converting gain, K_p	$2.79\ 10^{-5}$ [m/rad]
Operator stiffness, K_{op}	500 [N/m]

It should be noted that these operator's dynamic parameters may change during operation. For simplicity, the first two terms on the right hand side of (15.6) are omitted, and K_{op} is taken as a constant.

The constants of the parameters of the telehandshaking system are shown in table 15.1.

Since the master and slave are assumed to be physically equivalent, all parameters of both VIs for the master and slave should be the same. Each VI is represented by the following transfer function.

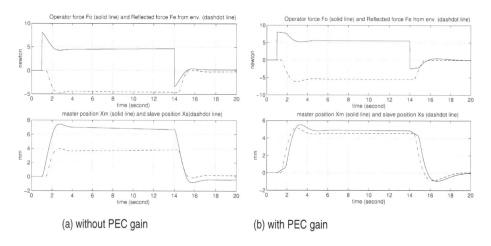

(a) without PEC gain   (b) with PEC gain

Figure 15.12: Simulation results of VIPEC with time delay (time-delay : 400 [ms]).

$$V(s) = \frac{1}{M_v s^2 + D_v s + K_v} \qquad (15.7)$$

Simulation

It is assumed that the master and the slave are physically equivalent, thus the parameters of both VIs should be the same. An operator tried to pull the spring by moving the master arm, holding it for a while, and then releasing it.

Similar to the above experiment, simulation of the teleoperator with time delay was performed in two cases: (1) VI without PEC (PEC gain = 0) and (2) VIPEC. In the simulation, the force exerted by an operator's muscles (F_{op}) was assumed to be a step function from first second to fourteenth second. The time delay of 400 ms was inserted in data transmission. The values of the parameters for simulation are shown in table 15.1. The parameters of both VIs, M_v = 10[kg], D_v = 1000 [N/(m/s)], and K_v = 20 [N/m] are the same. PEC gain was set to ten.

Figure 15.12 shows the simulation results for the system with 400 [ms] of time delay. In both cases, with and without PEC, the system is stable and force sensation is realized; however, VIPEC succeeds in reducing the position error between the master and slave positions.

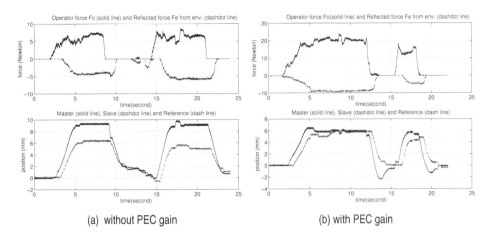

(a) without PEC gain   (b) with PEC gain

Figure 15.13: Experimental result of teleoperation

15.3.4 Experimental Results

The experimental results of the teleoperator with 400 ms (Figure 15.13) of time-delay agree well with the simulation results. The control approach, VIPEC, can reduce the position error between the master position and the slave position. As delay time increases, however, so does the position error between the master and slave. Here, the dynamics and the input forces to the both VIs are the same. Time delay (400 ms) is made in a computer.

Experimental data of the telehandshaking are shown in figure 15.14. For simplicity of implementation, two HSDs and two host computers were installed in the laboratory. The data from two host computers were communicated through the Internet via a server workstation which was located at Osaka University. Each host computer sent the data to the server workstation every 120[ms]. The server forwarded the written data to the other host computer every 100[ms].

The data were composed of three float numbers: the force applied by the operator, the force reflected from the environment, and the slave position. Therefore 12 bytes of data were transmitted through the Internet at each time.

The VIPEC was implemented on the HSD. The parameters of the VI were $M_v = 10$ [kg], $D_v = 800$ [N/(m/s)], and $K_v = 100$ [N/m]. PEC gain was two. VIPEC realized in the host computers ran at a 250 [Hz] interrupt rate.

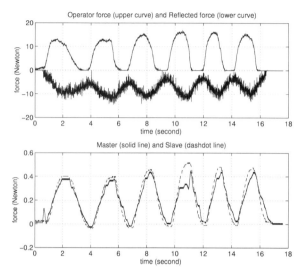

Figure 15.14: Tele-handshake between Tokyo and Osaka.

Two operators succeeded in shaking hands with each other and achieved force sensation through the HSD with VIPEC control approach. The position difference between the master and slave positions was small.

15.4   Conclusions and Future Work

This chapter discussed two experimental systems of Robotic Network System (RNS) to show the possibility of computer networked robotics. The first system, the mobile robot MUSEAR, was controlled by an operator at a remote host. An operator was watching a color video image transmitted through the Internet during teledrive. In this experiment, a robot was playing as a physical agent from an information space. The second system, HandShake Device (HSD), was also shown as a physical agent. A force information was displayed, which was physical information through the network, to an operator by using a network and a robot technology. A robot, especially a haptic interface (a master system of teleoperation) like HSD, is defined as a display for new media such as a force. That experiment used a time-delay compensation which is proposed as VIPEC, and showed good results of telehandshaking between Tokyo and Osaka. These experimental systems were run under a variable time-delay that caused the instability of the systems, which has to be solved as future work.

References

[1]   E. Paulos and J. Canny. Delivering Real Reality to the World Wide Web via Telerobotics. *Proc. of IEEE Int. Conf. on Robotics and Automation*, pp. 1694–1699, 1996.

[2]   K. Goldberg, M. Mascha, S. Gentner, et al.. Desktop Teleoperation via the World Wide Web. *Proc. of IEEE Int. Conf. on Robotics and Automation*, pp. 654–659, 1995.

[3]   W. J. Book, H. Lane, L. J. Love, et al.. A Novel Teleoperated Long-Reach Manipulator Testbed and its Remote Capabilities via the Internet. *Proc. of IEEE Int. Conf. on Robotics and Automation*, pp. 1036–1041, 1996.

[4]   Y. Kunii and H. Hashimoto. Tele-Handshake Using HandShake Device. *Proc. of IECON 95* 1, 1995 pp. 179–182.

[5]   WITTENSTEIN Motion Control GmbH: Roller screws, pp.5–7.

[6]   S. Nojima and H. Hashimoto. Master-Slave System With Force Feedback Based on Dynamics of Virtual Model. *Proc. of ISMCR'94: Topical Workshop on Virtual Reality*. pp. 93–100, 1994.

[7]   K. Furuta, K. Kosuge, Y. Shinote, and H. Hatano. Master-slave manipulator based on virtual internal model following control concept. *Proc. of IEEE Int. Conf. on Robotics and Automation*, 1987, pp. 567–572.

[8]   M. Otsuka, N. Matsumoto, T. Idogaki, K. Kosuge, and T. Itoh. Bilateral telemanipulator system with communication time delay based on force-sum-driven virtual internal model. *Proc. of IEEE Int. Conf. on Robotics and Automation*, 1995, pp. 344–350.

[9]   M. Suradechb and H.Hashimoto. Tele-Handshake Interface Based on Teleoperation with Time Delay. *Proc. 7th Int. Power Electronics Motion Control Conference and Exhibition*, 1996.

16 One Year of Puma Painting

Matthew R. Stein

16.1   Introduction

The PumaPaint project (http://yugo.mme.wilkes.edu/~villanov) is a web site allowing any user with a Java compatible web browser to control a PUMA 760 robot located at Wilkes University, Wilkes-Barre, Pennsylvania. Since the pioneers of online robots first opened sites in 1994, many other online robots have been introduced, and currently the NASA Space Telerobotics Program web site lists over twenty "real robots on the web." With these precedents, one may ask what is new about the PumaPaint site, what contribution does it make, and what motivated the author to produce it?

The PumaPaint site was opened to the public on 3 June 1998 after approximately three years of development effort, reported elsewhere [1, 2]. A conscious decision was made to avoid duplication of the prominent existing robot web sites, the USC Mercury Project [3], Australia's telerobot on the Web [4], the Telegarden (http://telegarden.aec.at/), and Xavier (http://www.cs.cmu.edu/~Xavier/). At the time these sites utilized what is called here a "set-submit" cycle where the user sets input parameters on a web form, then activates some type of submit button and waits for a response. Activation of this button would transmit the information to the server usually by activating a Common Gateway Interface (CGI) script. The response would often be a new camera view and display of updated status and position information.

The "set-submit" cycle is functionally equivalent to the "move-and-wait" cycle that is well documented in the field of time-delayed teleoperation [5]. Early researchers found that a user trying to control a manipulator through an interceding time-delay will naturally adopt a "move-and-wait" strategy where the user makes a conservative move and waits for feedback before making the next move. Other authors have shown that this strategy will increase the time required for completion of a task in proportion to the number of distinct motions times the round-trip time-delay. Thus, even moderate delays of 5 to 10 seconds can dramatically increase task completion.

The developers of PumaPaint adopted a Teleprogramming [6] approach, previously employed (but not invented) by the author. Teleprogramming has been

The Teleprogramming Concept

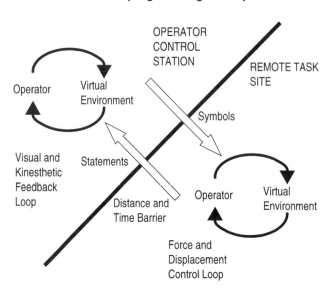

Figure 16.1:   The teleprogramming concept.

shown by this and other authors to potentially reduce task-completion times by reducing the amount of time waiting for feedback that is symptomatic of the "move-and-wait" strategy. Figure 16.1, from [7], presents the teleprogramming concept. Operators interact with a virtual representation of the remote site, and from that interaction, commands are sent across a distance and time barrier to the robot for execution. At its theoretical limit, the completion time of the remote task will be one round-trip communication delay longer than the completion time if performed locally. Deviation from this theoretical optimum will always occur due to a mismatch between the virtual and real environments and/or the inability of the remote robot to execute the commands. To the extent that the robot can perform the desired commands based on virtual interactions, however, the teleprogramming approach can provide a significant improvement over the "mov-and-wait" strategy.

Because CGI scripts limited web-based user interaction to a relatively small set of fields, menus and forms, and a single "submit" action, the virtual interaction needed for a teleprogramming interface simply was not possible. The Java programming language was newly released during the early development phase of

PumaPaint. Java programs are executed (interpretively) directly on the user's machine, allowing real-time closure of the operator-virtual environment loop shown in figure 16.1. The choice of Java was essential for a teleprogramming approach to online robotics and this distinguished the PumaPaint site from the existing online robots at the time.

16.2   Design

Rather than developing new features, much of the development work on the robot motion server involved improving the robustness, that is, the tolerance to variations from normal program behavior. Common events include disconnection at arbitrary times in a session, broken pipes, and incomplete packets. Also, some proxy-servers (e.g., AOL) stay connected long after the user has apparently lost interest in painting. A 20 minute inactivity timeout was also necessary to forceably disconnect these servers.

16.2.1 Software Architecture and Design

The robot motion server was developed prior to the inception of the PumaPaint project for the purposes of allowing collaborative use of the robot over the Internet. The motion server was designed with the concept of a daemon serving robot motions to Internet clients connecting on a TCP/IP socket. When a connection is established, the daemon reads and interprets 188 byte fixed-size packets from the socket.

On receipt by the motion server, the packet is checked for validity then forwarded to and immediately executed by the robot controller. When the command is completed, a response packet indicating acknowledgment and status is returned. The small set of commands available implements perhaps the most trivial form of robot programming language. Packets can specify joint and Cartesian motions, gripper actions, and initiate preprogrammed motions (macros). No sense of time is built into the motion server and thus all moves occur at the default speed of the robot. Destinations are specified as homogeneous transforms relative to a home position (10cm back from the center of the easel).

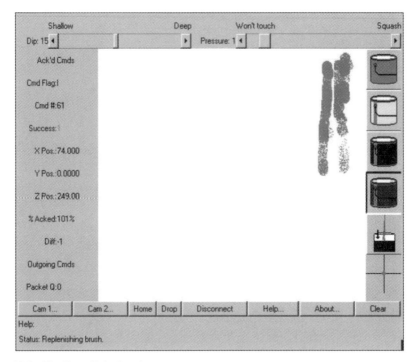

Figure 16.2: The PumaPaint interface.

16.2.2 Interface Design

Because the Java applet is executed on a web user's machine, the user interacts directly with the applet and receives immediate feedback. Figure 16.2 shows the interface as a user would see it within a browser window. The center portion is a virtual canvas and the main area of interaction. By clicking, holding, and dragging the mouse in this area, the user issues commands to the remote robot to apply paint to the real canvas. These mouse actions also cause the selected color to appear on the virtual canvas. Simply turning pixels beneath the mouse to a selected color however would mislead rather than assist the user, so several features were added to attempt to increase the fidelity of the virtual canvas. As shown in figure 16.2, the virtual canvas is colored as a blob rather than a shape with sharply defined edges. The blobs contain randomly generated gaps and streaks, and the proportion of area turned to the selected color progressively decreases as the brush stroke continues.

Figure 16.3: Camera 1 view.   *Camera 2 view.*

The bottom panel contains two buttons that spawn separate windows for cameras 1 and 2 shown in figure 16.3. Camera 1 is mounted on (and moves with) the robot, while camera 2 shows a "canvas shot" where the user can view the entire painting and about half of the moving robot.

All interface features attempt to aid the user unobtrusively. For example, virtual brush stroke may become clear to remind the user to replenish the paint, but no attempt is made to take this control away from the user and replenish automatically. The "pressure" can be set between numeric values labeled "won't touch" and "squash" as shown in figure 16.2. "Squash" will do just that, compressing the brush against the easel. Although this often results in ugly splotches of paint on the canvas and wear and tear on the brushes, the design does not prevent the user from doing this.

16.2.3 Robot Hardware

Project hardware consists of a PUMA 760 robot equipped with a parallel-fingered pneumatic gripper and controlled by a Sun4/110 workstation running RCCL [8]. The easel and all paint fixtures were constructed from inexpensive lumber and plastic as shown in figure 16.4. The paint brushes are the type frequently used by school children made by Crayola, Inc. Held in fixed positions beneath plastic funnels are four paint jars containing red, green, blue, and yellow, all Prang ready-to-use tempera paint. A homemade easel that holds a 30" x 50" counter top in a vertical position is shown in figure 16.4. Tonka toy treads are glued onto the

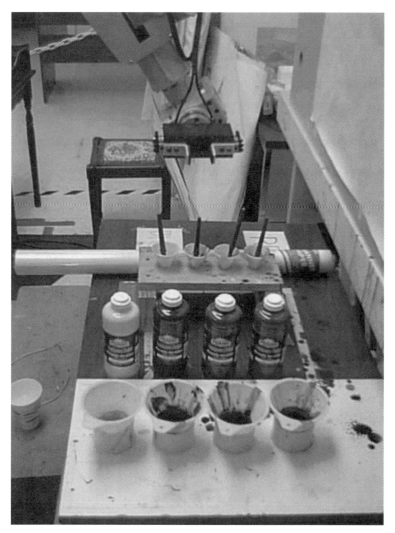

Figure 16.4: PumaPaint hardware.

parallel-fingered gripper for better grasp stability. Both the tabletop and the easel are suspended on leaf springs above contact switches so that excessive pressure or collision forces will trip the switches and cause immediate power cutoff. Little attempt has been made to ensure perfect reliability of the brush fixtures and paint jar holders constructed of handmade wood, glue, and duct tape assemblies that occasionally fail. The robot typically drops a paintbrush every five to six hours of usage and mishandles a brush with about the same frequency. Excessive contact forces on the table caused power cutoff about once every ten days.

16.2.4 Control

Despite the problems caused by inexpensive fixturing, the author has purposely declined to improve it. Devices already exist that allow the user to design graphics off-line and then produce perfect paper replicas — they are called printers. The purpose of the site is to engage the user in the act of remote painting, and it has attempted to avoid falling into the paradigm of "teleplotting" — the robot acting as a big (and imperfect) plotter. Rather, the site attempts to reinforce the user's role as an operator and the need to observe the task and compensate for mechanical imperfections inherent in most real-world applications. One such imperfection is manifested in the grasping of a paint brush handle by the parallel fingered gripper. Variations in the grasp produce marked differences in the resulting brush stroke, and if users seek to control the results, they need to account for this variation.

16.3   The Setup: PumaPaint

16.3.1 General Description

The author taped a fresh sheet of inexpensive banner paper on the easel two or more times daily, and cleaned the brushes and filled the paint jars before leaving each evening. The author typically arrived the next morning to find the canvas filled with paint and the brushes on the floor.

16.3.2 Experience and Statistics

The access log of the http daemon as analyzed using the program Analog [9]. Table 16.1 shows the number of downloads for the main PumaPaint HTML page, the Java interface, and the communications class. The later is only loaded after the user has entered a password and then selected a "connect to server" button. The distinct hosts downloading the communication class is the most conservative statistic and is probably the most accurate measure of the number of persons using PumaPaint. Based on this measure it appears that around 5,000 people have attempted to paint with a robot over the Web this year.

Table 16.1: PumaPaint download statistics for one year.

Hosts downloading PumaPaint components	Total	Distinct
Downloads of the main PumaPaint HTML page	18,039	11,813
Downloads of the interface	9,802	6,648
Downloads of the communications class	6,927	5,593
Average successful requests per day	18	
Domains outside the United States served	58	

Table 16.1 indicates there is about a 3,000 (or almost 30 percent) difference between total downloads of the interface and total downloads of the communication class. It is difficult to ascertain from the access log the source of this difference, but the following causes can be conjectured:

1. *The user is just "wandering through".* A user follows the link for the interface not genuinely understanding the site or the meaning of the link and then immediately changes his or her mind and presses the "back" button.

2. *The user tires of waiting.* The first significant delay using the PumaPaint site occurs when the interface is downloaded. The user's machine must start the Java interpreter and then load the classes necessary to execute the interface. Users may begin to download the interface intending to use it, but then they lose interest in the face of the download delay.

3. *The password window gets lost or closed.* The password window is a separate window that can be easily lost or closed while waiting for the download to complete. If this happens the user will see the entire interface but it will not work. Users also may get confused when confronted with a window that asks for a password and exit. Anecdotally, the most common e-mail message received from potential users is a request for a password. This despite the prominent (H2 font size) display of the public password on the main page.

4. *The browser is not capable of interpreting Java.* If the user's browser is not equipped with a Java interpreter, then following the interface link (clearly marked "requires Java") will produce a blank screen in the browser.

5. *The interface simply does not work.* This should not be overlooked as a potential cause of users not connecting to the robot, nor should the fact that numerous people have succeeded indicate that all users could. Because of the diversity of browsers, platforms, and Java implementations available, one

cannot be certain that the interface applet will work in all cases. Although there is no direct evidence of this, from this experience with Java one suspects that the interface is simply failing for a significant number of users.

One statistic (noted in table 16.1) is the average successful downloads of the communications class per day is eighteen, or less than one per hour. This statistic demonstrates the author's consistent observation: the robot sits idle most of the time. Anecdotally, the most frequently asked question by visitors to the lab is "what happens if two people try to connect at the same time?" The technical answer is that motion commands are queued and will be executed as soon as the current user releases the robot (not a very good answer–this is a useless feature), but the more relevant answer is that such occurrences are extremely rare. The vast majority of the time anyone in the world interested in doing so may take control of the robot with no waiting.

Between 800 and 1,000 machines downloaded the PumaPaint interface in a typical month. The site was most active in November (1,427 downloads) probably due to a favorable article in the *New York Times* on November 19, 1998 and subsequent distribution by the Associated Press (AP). Usage increased in March (1,049) probably due to an article in the *Chronicle of Higher Education* and mention in *Playboy* magazine. Access statistics also show the robot is typically about twice as busy in the late afternoon as it is overnight and about 50 percent busier on weekdays than on weekends.

Perhaps the most interesting aspect of the PumaPaint project is its ability to provide for the spontaneous creation of physical artifacts via the Internet. Given this uncommon capability, the most interesting question appears to be: "What did we do with it?" PumaPaint users produced 390 painted images in the first year of operation, and an examination of these images provides the answer to this question. In summary, the painted canvases can be divided into five major categories: nothing, text, icons, vandalism, and art.

Of the 390 canvases, 60 percent contain test strokes, spurious lines, squiggles, or other marks not identifiable as an attempt to create a meaningful image. This is often the result of leaving the canvas in place overnight when any number of users can access the site and overwrite or even tear down the works of previous users. During the day canvases were saved at user request or if the author thought they contained an interesting image.

Strokes that are recognizable as text appear on 51 percent of canvases. The

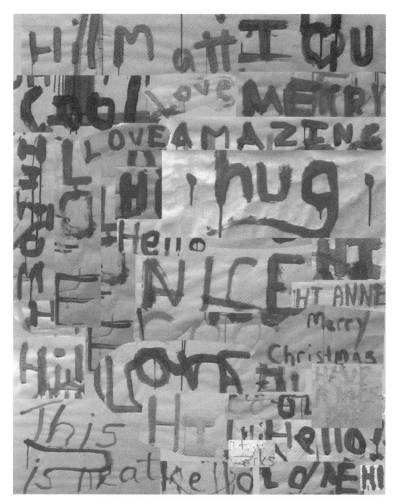

Figure 16.5:   Collage of text.

most common messages were self-identification (75), "hi" and "hello" (22), messages to others or with meaning unknown to the author (20), praise of the site (7), and requests for more paper or paint (4). As demonstrated by figure 16.5, a collage of text and messages created by PumaPaint users, the majority of messages were positive and friendly.

Many canvases have some paint strokes that are clearly identifiable as an attempt to create some form of painted image (i.e. something that is evidently not random strokes or text). This includes elaborate paintings as well as icons, small

Figure 16.6:   Common icons.

graphics, and geometrics. The smiley-face was the most frequent icon (25) followed by the heart (16). Simple geometrics (squares, 3-D cube renderings, triangles, cylinders, etc.) appear on 25 canvases, while elements of nature (sun, flowers, trees, etc.) appear on 21 canvases. Figures 16.6 and 16.7 show example collages of these images taken from multiple canvases created throughout the year. These images are of interest not so much because they were particularly successful or beautiful but because they were clearly intentional. Of all the canvases, 229 or

Figure 16.7:   Common schema.

59 percent contain evidence of some deliberate attempt to create a meaningful graphic.

Some users sought to vandalize the canvas and to do so was entirely possible by moving the pressure setting to "squash" and repeatedly going over the same area. The apparent goal was to rip off the paper and paint on the easel, as shown in figure 16.8. It is somewhat difficult to count deliberate acts of vandalism because the robot has enough mishaps on its own. The "blooper-reel" portion of the PumaPaint site shows thirty different incidents of brush or paper mishandling. The

Figure 16.8: Vandalism.

robot is entirely capable of doing this without user intention, and there was conclusive evidence of vandalism in only about five cases. Thus, although vandalism was an option that some users chose, it was far from prevalent.

The "Hall of Fame" section of the site contains fifty images that the author considers deserving of special attention and figure 16.9 is a collage of some of these images. Although land and seascapes are often attempted, the author found the greatest appeal in portraits. Not included in this fifty are some other very fine images that were unfortunately obscured or marred by subsequent users. If these images are included, then nearly 20 percent of the canvases contain some form of artwork conceived by the user and executed remotely over the Internet.

An offer to mail canvases to users free of charge is featured prominently on the main web page. About 50 users requested their paintings, 80 percent of which were mailed within the United States and the remaining to western Europe, Israel, and Australia. As would be expected, about half of "Hall of Fame" canvases were requested by their creators, accounting for about half of all canvases sent. Interestingly, the other half of mailed canvases, containing little identifiable and resembling graffiti to the author, were nonetheless requested and eagerly received by users.

Figure 16.9:   Collage of finest images

16.4   Conclusions and Future Work

Despite the precedent online robotics sites, the author expected that this site would also be of interest, perhaps of more interest, because of the carefully chosen task of painting and the ability to physically receive the artwork. Although some results were anticipated by the author, unanticipated results are likely more interesting and informative. The outcomes listed below went directly counter to the author's expectation or were results about which the author had no foreknowledge or expectation.

1. *The lack of commitment demonstrated by most users.* The majority of users gain control for a short while, make a few trial strokes, and depart without comment. This is symptomatic of the lack of commitment shown by web users reported more extensively by other authors [10].

2. *Infrequent vulgarity and vandal/hacker attacks.* The author was completely surprised by the apparent absence of hacker-cracker attacks on the site. It was anticipated that there would be an strong appeal to potential hackers to wreak physical havoc with a robot connected to the web, and a good deal of time as spent protecting against this. No known attempt has been made to compromise these precautions. It is easy enough to vandalize the site with the controls provided and although up to thirty users chose to do this, this number is much smaller than the author's expectation. One quite unexpected outcome is of these 201 canvases with text, only 7 contain text that is clearly identifiable as profanity or vulgarity.

3. *The lack of interest in controls.* For the most part users simply accepted the default settings for pressure, dip depth, and image quality despite their effect on the resulting painting. In spite of an extensive, hypertext-like help system, the author can conclude that for this site, like many other items of technology, users only read the instructions when all else failed.

4. *The lack of interest in images.* Despite a prominent message on the main page begging users to look at the camera images before painting, a sizeable minority of users never downloaded a single camera image. It was anticipated that witnessing the robot perform commanded motions "in real life" would be the main fascination with the site, but a good portion of users had no such interest or did not understand that this was possible.

However modest is this achievement of creating artwork remotely, it does demonstrate the fundamental concepts of teleprogramming. Whether the artwork is good or bad, it is without a doubt the result of intentional control of a distant operator via the Internet. In addition, the PumaPaint interface allowed the operator to create masterpieces (or junk) without adopting a visual-feedback based move-and-wait strategy. Operators interacted with a virtual interface and received immediate virtual feedback while this interaction caused a series of commands to be generated and later executed by a remote robot.

The primary interest in the project for the author, and the motivation for developing it, was curiosity as to what the world at large would do with a blank canvas and the ability to fill it. After a year of operation the author has in hand a body of evidence that provides the answer–mostly nothing. Not particularly worthwhile or interesting images comprise the majority of canvases produced.

It is easy to think of improvements to the interface or the experimental setup that would make the site run better. Often suggested are automatic paper changing, integrate video feedback directly into the virtual canvas, speed up video images with dedicated hardware, add features to the interface or implement it using new technology (e.g., plugins). Although it would be fun to add features to the site, the author has paused to ask several questions. Why implement a system that treats canvases with care when so few users request them? Why spend time putting in features that not one user out of 5,000 has asked for? The interface already provides the opportunity to achieve remarkable subtlety, greatly affecting the resulting visual impression. A year of operation has produced a sizeable bulk of direct physical evidence as to what users were capable of achieving. Continued development of the PumaPaint project will most likely address the following proposition: if we are more clever about creating an interface, will the art improve?

Acknowledgments

The initial concept of painting with the robot was developed in conjunction with Dr. Karen Sutherland, while at the University of Wisconsin, La Crosse. Mr. Chris Ratchford, a student of Wilkes University, was responsible installing RCCL and developing the initial robot motion server. Pete DePasquale developed the Java interface as a master's project at Villanova University working with Dr. John Lewis. I am extremely grateful to all of the individuals cited, and also to Wilkes University for providing the robot, laboratory facilities, and the postage for mailing the canvases.

References

[1] M.R. Stein, C. Ratchford, K Sutherland, and D. Robaczewski. "Internet Robotics: An Educational Experiment." *Telemanipulator and Telepresence Technologies II*. SPIE, vol. 2590, 1995.

[2] P. DePasquale, J. Lewis, and M. Stein. "A Java interface for asserting interactive telerobotic control." *Telemanipulator and Telepresence Technologies IV*. SPIE, vol. 3206, 1997.

[3] K. Goldberg, M. Mascha, S. Getner, and N. Rothenberg. "Desktop Teleoperation via the World Wide Web." *IEEE International Conference on Robotics and Automation*, Nagoya, Japan, 1995

[4] K. Taylor and J. Trevelyan. "Australia's Telerobot On The Web." *26th International Symposium on Industrial Robots*, Singapore, Oct. 1995.

[5] W. R. Ferrell and T. B. Sheridan. "Supervisory Control of Remote Manipulation." *IEEE Spectrum,* vol. 4, no. 10, 1967.

[6] R P. Paul, C. P. Sayers and M. R. Stein. "The Theory of Teleprogramming." *Journal of the Robotics Society of Japan*, vol. 11, no. 6, 1993.

[7] M. R. Stein. "Behavior-Based Control For Time Delayed Teleoperation." *Ph.D. thesis*, University of Pennsylvania MS-CIS-94-43.

[8] J. Lloyd and V. Hayward. "Real-time trajectory generation in Multi-RCCL." *Journal of Robotic Systems*, vol. 10, no. 3, pp. 369–390.

[9] Analog is a copyright (C) of Stephen R. E. Turner and is freeware available at http:/www.statslab.cam.ac.uk/~sret1/.

[10] K. Taylor and B. Dalton. "Issues in Internet Telerobotics." *International Conference on Field and Service Robotics*, Canberra, Australia, December 1997.

17 Online Robots and the Remote Museum Experience

Steven B. Goldberg and George A. Bekey

17.1   Introduction

The World Wide Web is primarily a network for access to and distribution of information in the form of text or audio or video. It is only during the past several years that a number of researchers have become interested in using the Internet to interact with the physical world. This is a fascinating extension of the network, from being primarily a brain-to-brain connection to also being a medium for muscle-to-object connections. Thus, the Internet has been used to remotely control Khepera mobile robots in Switzerland, a telescope in Australia, a guide robot in a museum in Pittsburgh, and a robotic garden originally at USC and now in Austria, to cite only a few examples. Such remote controls of physical devices are classical cases of "action at a distance," which in the past have been reserved to magicians or psychics capable of performing feats of telekinesis.

It should be noted that the projects described here are properly in the domain of "telerobotics," since they concern control of objects or systems by humans at a distance. Telerobotics is a well-known technology in the space program [1] and in the nuclear industry, where it is necessary to exert human control at a distance.

17.1.1 Background

At the University of Southern California the authors have been interested in Internet-based telerobotic systems for a number of years. The first project, known as Project Mercury [2], allowed remote users to control a robot arm positioned over a circular tray filled with sand, in which small objects were buried. At the end-point the robot was equipped with a compressed air nozzle and a camera. After positioning the robot over a desired spot, as seen on a remote user's monitor, the nozzle could be remotely activated using the mouse. The resulting blast of air would move the sand, and excavate to find the buried treasures. Successive blasts of air tended to move the sand from neighboring locations and rebury the objects. The second experiment, the Telegarden [3], allows use of a robot arm to view and manipulate a remote garden filled with living plants. (Project Mercury is no longer available, and the Telegarden was on display in Linz, Austria, as part of the Ars Electronica art exhibit, http://telegarden.aec.at until September 1999). It provides a monocular view of the garden and allows remote users to select the view location, plant seeds, and water the plants. The Telegarden's interface is a top projection of the robot and a zoom slider, offering only three degrees of freedom. It uses images returned by a CGI script on the associated server. Both of these projects received a large number of visits and became known internationally. The Mercury Project proved enormously popular, registering nearly two million hits over a period of seventeen months. The Telegarden was active at USC for about one year, during which time it received about 500,000 hits. During summer 1996 it was moved to Linz, where it continued to receive international attention. The Telegarden won the Kobe prize at the Interactive Media Festival and won first prize at the Festival for Interactive Arts (FIVA).

The remainder of this chapter we describes DIGIMUSE, an interactive telerobotic system allowing remote users to visit an art museum and inspect a three-dimensional sculpture. It then discuss the issues involved in adding the sense of touch to this experience as a further extension of remote contact with physical objects.

17.1.2 Goals of the DIGIMUSE Project

The primary goal of this project was to provide a means by which people not able to visit a sculpture gallery could study pieces in its collection. In the past, only two-dimensional objects of art have been accessible through the Internet — paintings, drawings, and the like — along with photographs of three-dimensional objects. To experience the depth of statues or other 3-D objects, it was necessary to be physically present to view the object from different positions and ultimately to touch the object. This limitation has been partially overcome and an entire realm of exhibition space has been opened up by creating an environment in which the object can be viewed stereoscopically from virtually any angle and from anywhere in the world.

A museum's exhibition space is always limited, both by the lack of physical space for the exhibits and the limited access available to visitors. The University of Southern California's Fisher Gallery is an accredited art museum. In common with most other museums and galleries, the Fisher Gallery can display only a small part of its collection because of a lack of physical space. Hence museums rotate large parts of their collections, leaving only smaller permanent collections on display at all times. If a portion of the physical space were devoted to creating a virtual exhibition space, museums may be able to make better use of the available facilities.

Computer and Internet technology can be used to create virtual exhibits that augment the physical exhibit space as well as allowing remote viewing of the space. It is impractical for an art student in Berlin to fly to Los Angeles to view one statue. If a web page gives a student the ability to view all aspects of the statue in stereo, however, then the limitations of access and proximity are removed.

There are two ways in which this extension of the physical space can be achieved. The first requires photographing all the objects to be exhibited. For 3-D objects this requires that they be photographed from a number of orientations, and that all these images be stored and indexed appropriately, thus creating an electronically accessible "virtual museum," the size of which is limited only by the available storage on server computers. The second approach allows for interactive viewing of a single work of art in real time, using a robotic system for positioning and orienting both the object and the cameras. Clearly, the latter approach is limited to viewing one object, but it does offer the viewing visitor the sense of participation that arises from the real-time interactive features. The second of these approaches has been chosen for the project described here. In both approaches the work of art can be viewed from anywhere in the world and at any time of day or night.

This project, called DIGIMUSE (for "digital museum") is an extension of the two previous experiments in teleoperation over the Internet developed at USC, as described in the introduction.

17.1.3 Issues in Telerobotic Systems

In common with other robotic systems, Internet-based telerobotic systems require consideration of:

- Sensing

- Actuation

- Computation

The initial design of the system described her was based entirely on the use of cameras to provide visual information to the user. As described later in the chapter, a tactile sensing interface is currently being designed. Actuation for this system requires the ability to position the cameras and/or the sculpture so that the desired

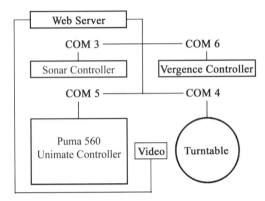

Figure 17.1: Major components.

views can be obtained. Connecting the sensor inputs and actuators is a computer used to interpret and translate the user's intent into camera position and to provide the signals needed for visual feedback.

17.2   System Design

The major components of this system consist of a six degree of freedom robot arm (which holds the cameras), a rotating platform (which supports the statue being viewed), computers and the associated electronics and network connections as shown in the block diagram of figure 17.1. The robot and turntable occupy separate workspaces. The workspaces were intentionally separated to make it impossible for the robot to come into contact with any part of the statue. It is important to note that the complete installation is located within the art gallery, thus becoming an exhibit in its own right. Plexiglas partitions separate the robot from the live audience for reasons of safety. This project used a sculpture named "Trinkendes Mädchen" (Drinking Maiden), by German sculptor Ernst Wenck, which was placed on the rotating platform at USC's Fisher Gallery.

The end-effector of the robot supports a vergence that which aims the two cameras as a function of the distance to the object (measured by a sonar proximity sensor), thus assuring the correct viewing angles for stereo images. The two CCD cameras are mounted under the convergence head to reduce the singularities created by the end-effector. Each CCD camera has autofocus and a computer-

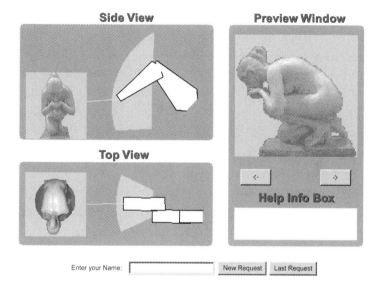

Figure 17.2: User interface.

controlled powered zoom which, when implemented, will allow the user to get a closer view of the statue than the robot's workspace will allow.

17.2.1 Interface Design

The user interface provides a three-image, six-click control of the manipulator. The first image is a side view of the statue and robot arm, shown in the upper left portion of figure 17.2. In this view the user chooses the camera aim point on the statue in the x-z plane. By clicking in the light gray, the user can position the camera in this plane. A thin yellow line in the display shows the aim of the cameras.

The second image is a top view of the statue and robot arm which allows the user to control the pan of the cameras. The positioning method is the same. To uniquely specify the aim point of the camera on the statue, the user selects the location and orientation in both planes. The third image (labeled "preview window" in figure 17.2) allows the user to see (approximately) what he or she will see when requesting a particular statue orientation and camera orientation. In this view the user chooses the orientation of the statue. To minimize the need for actual rotation of the turntable, photographs of the statue have been stored which were taken at 45-degree intervals about the z-axis. Each click on the arrows below the image results in a clockwise or counterclockwise 45-degree rotation of the virtual statue. This image is intentionally locked into 45-degree perspectives to inspire the user to use the robot arm to obtain a particular image, which is more precisely aligned with the user's desires.

By clicking the "submit" icon at the bottom of the screen, a command is sent to the server where it is queued. When that request reaches the highest priority, the turntable rotates and the robot moves to the desired position. The pictures are taken, compressed and sent back to the remote user as JPEG images. The users can see a stereo image using both eyes when using the head-mounted display (HMD). The system currently supports field sequential display for virtual IO HMDs. For users who do not have HMDs, the system provides two images side by side.

In contrast with the three degrees of freedom available to the Telegarden, the museum interface allows the user to manipulate a pair of cameras in six degrees of freedom as well as rotating the statue, thus obtaining a total of seven degrees of freedom. By using a Java applet, the load required by the server has been lightened and the interface has been made interactive.

17.2.2 Robotic Hardware

The arm is a Puma 560 robot, which is controlled by a modified Unimate controller over a serial line. The controller runs a sequence of VAL programs, which implement the serial commands. The main program accepts a list of six numbers to specify the position and orientation of the end effector. These programs also limit the range of motion for the arm to keep it from damaging the statue or itself.

The end-effector of the robot is a TRC Zebra vergence head, which holds the two video cameras and a sonar proximity sensor. It measures the distance to the object being viewed and provides the necessary convergence of the two to allow for comfortable viewing of stereo images. The camera mounts are located 10 inches apart, which is wider than the distance between the human eyes. The closest focus point of the cameras, however, puts the virtual image at the point of convergence. A TRC servocontroller is used to interface the serial port with the vergence head.

The cameras are made by Mitsubishi (Model CCD-300U). They have two outputs, both of which provide S-video and composite NTSC signals and one RS-232 serial port. Both composite outputs are connected to a Coreco Oculus TCI-SE frame grabber and the S-video outputs are connected to an HMD for local, real-time viewing. The serial connection can be used to control the zoom. A single sonar panel is attached to the front of the Zebra head to measure the distance to the statue, which is needed to command the convergence angle of the cameras. This sonar panel is controlled by a TRC Sonar Ranger Control board via a serial port. The TCI-SE framegrabber can take pictures sequentially from up to three cameras. Each picture can have a resolution of 160 x 120, 320 x 240, or 640 x 480 pixels at 24-bit color depth. The card uses bus-mastering technology to pipe the image information directly to the video card on the computer. Hence, it was necessary to use a PCI-compatible card capable of handling the video data; a Diamond Stealth 3240 video card was used for this purpose.

The marble statue (which weighs some 800 pounds) is placed on an automotive turntable from the Chase Foundry capable of supporting 1,500 livres. An AC motor, geared down to turn at one rpm, drives the table. There are two starting windings, allowing for motion in either direction. The table is connected to the server by a custom controller designed and fabricated in the laboratory. The controller receives two commands, that determine the direction of rotation and engage the motor, respectively.

The server is a dual-pentium computer running Windows NT Server and Netscape's Enterprise Server. It is equipped with 32 mbytes of RAM, 2 gbytes of storage, a 3Com 10BaseT ethernet card and a four-port serial card.

17.2.3 Program Architecture

The major elements of the software architecture are shown in block diagram form in figure 17.3. When a request for pictures to be taken arrives from a remote user, it initiates execution of a local program on the web server which will either begin decoding the position data sent by the user's applet, or, if the system is busy, put the user in a queue.

Note that in addition to the turntable position (which can be specified in 45-degree increments), the user can command five degrees of freedom of the robot end-effector (three Cartesian coordinates: x, y, z and the rotational positions: pitch and yaw). There are only five because the cameras are not allowed to roll. Position data are stored in an environmental variable as a sequence of fields, which are delineated by slash marks. Once the variable is read, the string is broken down into

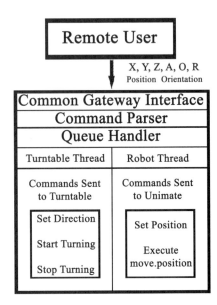

Figure 17.3: Software architecture.

individual commands. The first part is a two-letter command that either requests a new position or a direction in which to move. If a new position is requested, the next five commands are the required linear and rotational coordinate values. The next command is rotation of the turntable, which is specified in degrees, at 45-degree increments. Lastly, zoom is stored as one of four preset values. These numbers correspond to values returned by the cameras when queried for the zoom value.

Once the variables have been deciphered, the program spawns off three threads. The first thread controls the robot motion, the second rotates the platform, and the third sets the camera zoom values. All devices, with the exception of the platform, are controlled via a serial port. This posed a problem since a standard PC only has two or four serial ports, thus solution was to use four additional serial ports. Each port is given only its base port address so that all devices can run at the same time. After each thread finishes, the program requests the distance to the statue from the sonar and commands the vergence head to converge the cameras at that distance. There is sufficient overlap between the right and left pictures so that when viewed by each eye separately the pictures appear in stereo.

When all motion is completed, an image is taken from both cameras and saved as the location and orientation of all components. For example, if the robot is at *x,y,z* coordinates of 120, 689, 500 units and the angular orientation of the camera is given by 320 and 160 degrees, and the turntable is at position 15, the left image will be saved as L12068950032016015.jpg. This detailed instruction allows a user to reload the same image without moving the robot. Each image is removed after twelve hours to minimize disk usage. After each image has been taken, the program allows the next user to start the process. The current process returns html source code that points to a second Java applet whose parameters are the names of both images plus a third interlaced image which is used by several HMDs. This second image is created in another thread while the HTML code is being sent. This process may change to having the Java applet composite the image if the server is overloaded by the composite requests. The existing Java viewer shows both images at one-half resolution and allows users to select the type of HMD they wish to use to view the statue in a submersive environment. Initially, each image is sent for the side-by-side view windows. While these are being sent, the system creates the interlaced image.

17.3   Results

The system was capable of executing 3,600 complete access operations per day, including 15 seconds for the turntable, 5 seconds for the robot, and 5 seconds for image encoding. Not all hits, however, result in completed robotic operations. The system received some 250,000 hits during its first six months of operation after becoming public in February 1996 and it averaged 30,000 hits per month until it was decommissioned in July 1998.

17.4   Conclusions

The DIGIMUSE system is the third in a series of experiments performed at USC that allow remote users to control the motion of robots over the Internet. It makes possible the interactive viewing and study of statues and other three-dimensional art objects in real time, using only a computer, a modem, and a mouse. The addition of a head-mounted display makes it possible to view these objects in stereo. The system has been successfully demonstrated in an art museum, the Fisher Gallery, at the University of Southern California. It should also be noted that the system may also have important industrial applications, such as remote inspection of manufactured parts or components in hazardous or inaccessible environments. Current work is focused on adding the sense of touch to the telerobotic experience.

17.4.1 Future Work: Tactile Feedback

A further extension of the telerobotic concept will make it possible for a user to "touch" as well as to see a remote object. Two ways of implementing the tactile interface are being investigated. The first is based on the acquisition of haptic information by using Sensable's PHANToM force feedback robotic master and other devices, including a vibrotactile glove [4]. The PHANToM [5] is a small, desk-grounded robot that permits simulation of single fingertip contact with virtual objects through a thimble or stylus. It provides a means for remote visualization by enabling "touching" (tactile perception and discrimination) of objects and surface textures. Attention will first be devoted to the tactile appreciation of art objects by researching issues related to sensing their surface texture (feeling the local virtual terrain of the object beneath the fingers). The visitor to the exhibition will be able to "touch" the surfaces of selected objects from the collection via a force-reflecting device that exerts pressure on a stylus in the user's hand to simulate the experience of feeling surface texture. Surface texture data will be collected by a real-time, telerobotically controlled laser-point scanner. This method will make it possible to provide a synthetic but realistic sensation of touch, while protecting precious museum artifacts, since they would be laser scanned but not actually touched by a physical probe. It is also possible to use a vibrotactile display, which applies multiple small force vectors to the fingertips.

The PHANToM provides a point contact to a remote object. The use of special gloves equipped with vibrators will also be investigated, which will enable the user to sense when a virtual hand has made contact with an object. While not providing surface texture information, this method will make it possible to obtain the appropriate finger positions as one traces the outline of an object with the hand.

Acknowledgments

The development of the DIGIMUSE project was supported by an internal grant from the Annenberg Center for Communications at USC (principal investigators: Dr. Selma Holo, the director of the Fisher Gallery, Professor Margaret McLaughlin from the Annenberg School of Communication and Professor George Bekey from the Computer Science Department.) Dr. Holo allocated valuable exhibit space for the DIGIMUSE project, and allowed the use of the priceless "Drinking Maiden" statue for remote viewing. Professor McLaughlin and her students, along with the staff of the gallery, provided all the background on the statue as well as the text, art, and layout for the opening web pages. Professor Ken Goldberg (no relation to first author), currently at the University of California at Berkeley, was one of the originators of this project while he was on the faculty at USC. Since he was also responsible for the Telegarden and Mercury Project, we consider him to be the intellectual grandfather of this project as well. We also gratefully acknowledge sponsorship of the program by internal funds at SC.

References

[1]   S. Lee, G. A. Bekey, and A. K. Bejczy. Computer control of space-borne teleoperators with sensory feedback. *Proc. IEEE International Conference on Robotics and Automation,* pp. 205 – 214. St. Louis, Missouri, 1985.

[2]   M. Mascha, S. Gentner, N. Rothenberg, C. Sutter, and J. Wiegley. Desktop Teleoperation via the World Wide Web. *Proc. IEEE International Conference on Robotics and Automation,* Nagoya, Japan, May, 1995.

[3]   K. Goldberg, J. Santarromana, G. Bekey, S. Gentner, R. Morris, C. Sutter and J. Wiegley. A Telerobotic Garden on the World Wide Web. *SPIE Robotics and Machine Perception Newsletter* 5 (1), March 1996.

[4]   G. Burdea. *Force and touch feedback for virtual reality.* New York: John Wiley & Sons, 1996.

[5]   K. Salisbury and T. Massie. The PHANToM haptic interface. In *Proc. AAAI Spring Symposium Series, Toward Physical Interaction and Manipulation,* Stanford, CA, 1994.

18 Telerobotic Remote Handling of Protein Crystals via an Internet Link

Steven C. Venema and Blake Hannaford

18.1 Introduction and Literature Review

Growing crystals of biologically significant proteins in microgravity is a scientific opportunity unique to space. Protein crystals are necessary for x-ray crystallography, the major technique for determining the chemical structure of proteins. It is hoped that the use of microgravity will allow determination of the structure of new and medically significant proteins. So far no crystal has been grown in space that could not be grown in the earth's gravity field. Controlled studies have shown however, that crystals with fewer flaws can be grown in microgravity which leads to higher signal-to-noise ratio in the x-ray diffraction results [13, 15, 16].

There are concerns about the availability of crew time for experiments planned for the space station. Telerobotics may provide a solution to this problem by allowing some scientific operations to be conducted from earth by remote control. This chapter describes an initial demonstration of telerobotic capabilities for the handling of materials and samples in microgravity protein crystal growth experiments [7, 8].

Protein crystals grow in aqueous solutions under delicately controlled conditions in days or weeks. They are about 0.5 mm in size and of a gelatinous consistency. Before they can be analyzed by the x-ray beam, they must be mounted in glass capillaries of about 1 mm inside diameter. The process of mounting protein crystals is a delicate manual operation considered to be challenging for trained human operators. In one commonly used method, the crystal is first aspirated from its growth medium into the water filled capillary. Then water is removed from the capillary, causing the crystal to adhere to the capillary wall by surface tension, and the ends are sealed. This process is usually accomplished by a human operator looking through a dissecting microscope, and it is often performed in a cold room kept at 4 degrees centigrade.

The project described here was designed to determine whether it is feasible to perform the initial steps of protein crystal mounting via remote control over an internet link. The eventual goal is a functioning system aboard the space station in which ground-based scientific investigators can handle most or all of the protein crystal mounting steps by remote control.

Direct-drive (DD) robots have been studied for the past several years, primarily at a size scale similar to human arms [1, 11]. One of the key problems with DD robots is excessive power dissipation under gravity loads. Clearly this is not a problem in space. It turns out that even in the 1g environment, scaling analysis shows that as DD robots are made smaller, the excessive heat dissipation problem rapidly decreases [6]. Thus the benefits of DD, high precision, and high speed are available at this smaller scale for both earth and space based applications. The University of Washington's mini-DD robot is a small 5-axis manipulator designed to support precision manipulation tasks in biomedical applications. Engineering details of the robot have been previously reported [4, 5, 6]. It is ideal for the protein crystal mounting application because of its small size (about 15 cm x 30 cm x 10 cm), weight (800g moving parts), and power consumption (7 W nominal, 63 W peak). The small system size allows it to fit within existing space station facilities (such as the European Space Agency (ESA) microgravity glove box) with sufficient extra room for experimental equipment.

In the past few years, the Internet has been used for telerobotics. Kim and Bejczy [12] teleoperated a seven degree of freedom robot using a transcontinental Internet link for joint motion commands and a satellite link for returning video. Goldberg et al. [3] created a teleoperation user interface supporting point-to-point motion commands with still video returned to the operator using the web standard, hypertext transfer protocol (HTTP).

The space station communication link is the same TDRSS system used by the space shuttle; this link is expected to be a tightly contested resource during space station operation. Boeing Defense and Space is studying alternatives through which Internet links can be provided via one of the low earth orbit cellular satellite networks now under commercial development. This link could then provide Internet services to the space station.

Hirzinger et al. [9] performed the first telerobotic experiment flown in orbit. Their ROTEX experiment, which included control from the ground, confirmed the ability of an advanced, sensor rich arm to control contact forces and capture a free-floating object. The ROTEX arm was a full-sized robot arm however, and it was not integrated into an existing space research task.

Figure 18.1: Hardware configuration.

This chapter provides a high-level overview of the protein crystal growth demonstration system (PCGDS) designed by University of Washington (UW), Boeing Defense and Space, and Vanderbilt University primarily in 1995 and 1996.

18.2   Hardware Description

The system hardware can be divided into four groups: workspace, power control, manipulation, and video. The workspace (figure 18.1) is contained within a physical mock-up of the ESA space-station glovebox — a ventilated work area with a clear plastic front, approximately 0.9m wide by 0.6m high by 0.5m deep. Two ports in front allow an astronaut to insert gloves to directly manipulate an experiment. In the demonstration task, simulated protein crystals (0.5mm plastic cubes) must be captured and extracted from a drop of water in a 3 mm diameter well which is part of a 24-well CrysChem tray (standard biological labware used for growing protein crystals). The tray is situated on the stage of a microscope to provide a close-up view of the well, the crystal, and the robot end-effector.

Below the glovebox is an electronics bay containing power supplies, motor controllers, and power control relays. A programmable logic controller (PLCDirect series 205 Micro PLC) controls power for all subsystems. Both logical combinations of inputs and time sequencing can be programmed to control the state of the power distribution system using ladder logic. The PLC will initialize the entire system to a desired state after a power failure. Although the PLC is not envisioned as a component of a flight system, it provides a highly flexible and robust means to support the demonstration system.

Manipulation capability is provided by the UW five-axis mini-DD robot mounted on a linear motion base with up to 30cm of travel to extend the robot's reach to multiple work stations. The robot end-effector is a micropipetting system consisting of a capillary connected by flexible tubing to an automated syringe pump. The robot subsystem consists of the five-axis mini-robot itself, power amplifiers to run the robot actuators, and a control computer (TMS320C30 floating point digital signal processor (DSP)). The DSP board generates smooth trajectories between commanded end-points using a third order polynomial. PID control software for each joint generates actuator current commands for the power amplifiers. Control and trajectory generation operate at a clock controlled interrupt rate of 1000 Hz.

The linear motion base is actuated by a stepper motor, ball-screw, and controller which is capable of positioning the robot base to a precision of 25 microns. The syringe pump (KD Scientific, Model 210) displaces fluid in the capillary by precision displacement of a disposable syringe in an adjustable mount. Control for both the motion base and the syringe pump is provided by the server software (described below) through serial-port interfaces. A fluid valve controlled by the server and PLC and connected to a water reservoir can introduce fresh water into the tubing to purge the system of air bubbles.

Visual feedback to the operator comes from three video cameras mounted at different points in the workspace. The signals from these cameras are sent to a video switch for distribution. One of the cameras is a specialized microscope camera used to provide a detailed view of both the crystal position and pipette tip (figure 18.2). A second camera is attached to the moving robot base and provides a view of the robot manipulation workspace, and a third camera is set up so that the capillary can be positioned close to the lens to provide visual verification that the crystal has been captured. Another camera is wall-mounted for video conferencing between experimental teams at the two sites. The video switch routes NTSC video signals to a local monitor and to the network via three PC's with capture boards. Video signal routing can be controlled through commands from the server via a serial port on the video switch.

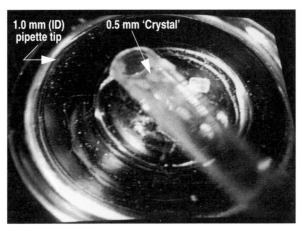

Figure 18.2: CU-SeeMe video from microscope camera showing simulated protein crystal and pipette tip held by robot.

18.3   System Overview

The glove-box system containing miniature robot, microscope, and cameras at the remote site (the University of Washington Biorobotics Laboratory in Seattle, WA) communicates with a user control station site (Boeing Defense and Space in Huntsville, AL). The two sites communicate via internet protocol (IP) (figure 18.3). The system was tested in two communication configurations, first, a private T1 line running Internet protocol between Huntsville and Seattle, and later the communication was switched to the commercial internet.

At the remote site, a server computer handles all commands and data with the exception of the video feeds. Video is handled by three identical 486 PC's with Creative Labs VideoBlaster video digitizer cards. The cards digitize NTSC standard video from the color cameras within the robot work cell using CU-SeeMe internet video conferencing software[1] [2]. CU-SeeMe sends each signal as a small (160 x 120 pixels), low-bandwidth (90kbps) color video stream over the Internet link to a computer running CU-SeeMe reflector software.

At the user control station, a laptop PC running CU-SeeMe provides video feedback to the operator from the three remote video sources. A separate PC

---

[1].   *See http://cu-seeme.cornell.edu and http://www.wpine.com/cu-seeme.html.*

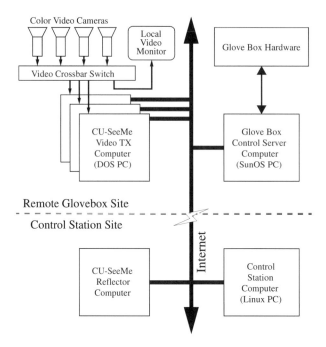

Figure 18.3: Communications topology.

running Linux and X-windows is used to maintain a menu-based user interface program. Remote display of this interface on the laptop PC allows both user interface controls and video feedback to be displayed on a single portable computer screen.

The user interface provides two distinct methods for controlling equipment at the remote site. The primary control method involves the use of predetermined macros that encode sequences of low-level commands such as "move robot along a trajectory to position X, move linear rail to position Y, and dispense Z microliters of fluid from the fluid pump." These macros are assigned to individual buttons on the graphical user interface, allowing the operator to quickly accomplish tasks by pressing a sequence of macrobuttons.

The user interface also allows low-level control of robot joints, linear rail position, fluid pump parameters, and soforth, via sliders and buttons in pop-up dialog boxes. This low-level control capability is intended only as a secondary or backup control method, and it is used for performing unplanned or experimental procedures as well as for generating new macrodefinitions.

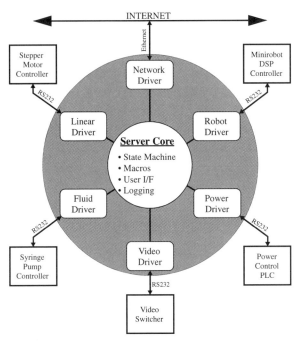

Figure 18.4: Server software structure.

Server Software

The glove box contains a number of devices that require external control. Each of these devices has its own specific communication protocol. To handle the specific requirements of multiple devices and coordinate activities between these devices, a generalized control program, which is called the server, was developed. The server has a core set of scheduling, logging, and command interface features that is coupled with a set of device drivers–one for each device (figure 18.4).

The driver software for each device consists of four functions with identical calling conventions:

open() —	Device initialization
close() —	Device shutdown
event() —	Parse incoming data from device
action() —	Send data to device

The event() routine passes device events such as "command completed, device error," or "data value X has changed" to a core state machine that is used to coordinate events and actions among multiple devices. Multidevice behaviors are encoded within this state machine as multiple sequences of event/action pairs. Each sequence has a pointer to the currently active event-action pair. An incoming event from a device causes the state machine to scan all currently active event-action pairs for any that are waiting for that particular event. Matching events cause the activation of the associated action (often an action on a different device) and the pointer is updated to the next event-action pair in that sequence. Provision is made for retrying actions at a later time should a given device action() routine return a code indicating that a later retry is desirable.

The software associated with each device (referred to as a device driver) maintains a standard data structure that contains information such as I/O file descriptors, pointers to action(), and event() functions, device status, and soforth. Pointers to each of these structures are returned by the device open() routine so that the server software may access these data as needed. Each device driver, however, is designed to be completely unaware of other devices.

The design described above allows all I/O channels from the server to external devices, including the network interface to the client, to be implemented as device drivers. This simplified the server program development and allows the addition of future devices with minimal difficulty at a later time.

A macrocapability is included in the server that allows sequences of device actions to be encoded and easily edited in a text file. The client library may be used to call these macros by name, providing the ability to activate multidevice coordinated actions on demand. This feature has proven to be extremely useful for rapid prototyping of tests and new activities. In operation, the macroinstructions related to the minirobot specify absolute or relative joint angle changes and a movement duration.

18.4   Communication Protocol

The communication protocol between the server and client software is designed to facilitate robot control in the presence of significant and variable delay. Although some delay is present in the terrestrial internet, the protocol is designed to function with low-earth-orbit links having delays in the range of 4 to 10 seconds round-trip. Several characteristics of the protocol were determined by this requirement. Because the standard TCP/IP communication protocol is ill-suited for this type of communication channel, a UDP/IP-based protocol was developed which better supports communication under these conditions.

The protocol uses datagram packets and implements time-outs and retries explicitly. Both command and link time-outs are implemented. Packets for each session are numbered sequentially to provide a check against packet loss and reordering. Old packets are discarded. A checksum in each packet guards against corruption.

All packets have a common eight-byte header consisting of a two-byte packet length, a two-byte checksum, a two-byte sequence number, a one-byte command, and a one-byte command argument. In addition, some packet types include command data or messages.

There are two components to the protocol: transaction sequences initiated by the client and a regular stream of status packets from the server to the client. The latter provide a connection check and also provide regular robot position updates to the client in the absence of user commands.

18.4.1 Client Library

The remote operator controls the system using a graphical user interface. The user interface was developed at Vanderbilt University's Measurement and Computing Systems Laboratory using their MultiGraph Architecture development tool (see http://mcsl.vuse.vanderbilt.edu) [17] and the Tcl/Tk graphics windowing development kit [14]. Due to space limitations, details of this user interface will not be discussed here.

The user interface is linked with a C-language client library that provides a high-level interface to the UDP/IP protocol. This library handles the details of connection setup, sending and receiving of packets, and determining command and connection status.

A key feature of the client library is that all commands return immediately, as the library keeps track of the command processing internally. This facilitates the library's use in single-threaded applications. The possible command status values are *unconnected*, *ready*, *sending*, *busy*, and *error*. Commands must be sent sequentially; a new command may be sent only when the previous command has finished. The initial status is *unconnected* until the connection to the server is confirmed. Then the typical status cycle is *ready* → *sending*→ *busy* (for commands that have a long execution time, like macros) → *ready*. The error status usually requires restarting the connection process.

The library API includes functions for global system management:

nr_connect() — Connect to remote server

nr_disconnect() — Disconnect from server

nr_setmode() — Set server control modes

nr_null() — Process data from server

The remainder of the functions can be grouped by device into four groups, "robot," "gripper", "auxiliary devices," and "parameters":

nr_set*() — Send data to server

nr_req*() — Request data from server

nr_read*() — Request local copy of data

where the* is substituted for the name of one of four device groups:

robot — Robot position data

grip — Gripper data

aux — Linear motion, fluid pump, video switch, etc.

parm — Control system parameters

So to set the robot position, one calls nr_setrobot(), while nr_reqrobot() will request a server copy of the robot position and control setpoint.

The nr_connect() function returns a pointer to the communication socket used on the client host. This allows the client to watch for incoming packets and to call a library function (typically nr_null()) to process them as needed.

System Operation

The glove box work cell demonstrates the capability to capture simulated protein crystals under interactive operator control using a series of macrocommands. The work cell is arranged as shown in figure 18.5. This demonstration task involves the following operations:

1. Initialization

2. Moving the robot-held pipette tip to the CrysChem cell under the microscope.

3. Hunting down and capturing the protein "crystal."

4. Moving the pipette to a close-up camera station to verify that the capture was completed successfully.

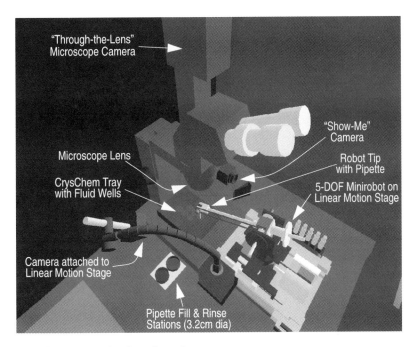

Figure 18.5: Glovebox work cell configuration.

To begin, the operator (in Huntsville, AL) first starts up the user-interface software, connects to the remote server computer over the Internet, and commands it to begin power-up sequencing of the glovebox work cell equipment (in Seattle, WA). This turns on lights, powers up the control systems and power amplifiers for the robot, linear motion rail, and syringe pump, and activates the small video cameras within the work cell.

Immediately following the power-up sequence, the state of glove box devices is not known. The minirobot uses incremental analog quadrature encoders for position sensing and must be calibrated to a known position after power-up. The linear motion rail and syringe pump both use open-loop stepper motors for position control so that they too must be initialized to a known position. A single macro allows the operator to completely initialize these devices to a known state.

The hydraulic tubing between the syringe pump and the pipette tip tends to develop large pockets of air in the line due either to evaporation out of the pipette tip or the emergence of dissolved gasses in the water. A single macro allows the user to initialize the fluid system by moving to a rinse station and flushing the entire line with fresh water.

Once all initialization is completed, the operator uses a GotoCell macro to safely move the pipette tip into the CrysChem cell that is being viewed under the microscope. With both the simulated protein crystal and the pipette tip visible in the microscope view (see figure 18.2), the operator can then guide the pipette tip down and over the protein "crystal" using macrobuttons that generate incremental ~250um motions along any of six directions relative to the camera view (up, down, left, right, in, and out).

Once the operator believes that the pipette is correctly positioned to capture the crystal, a "capture" macro is used to command the syringe pump to withdraw a controlled amount of fluid which will cause the crystal to enter the pipette. A "showme" macro then moves the robot and linear motion rail along a collision-free trajectory which safely positions the pipette tip in front of a video camera that displays the 0.5mm crystal within the pipette.

If the crystal capture operation failed (which happens more often during initial operator training), the operator can use additional macrocommands to move the pipette tip back to the CrysChem cell and re-attempt the capture.

18.5   Performance

A formal performance evaluation of the system has not been conducted. Several measurements can be reported however, to give a sense of what it is like to operate the system. The demonstration was implemented using macros which command movement times in the range of 0.5 to 2.0 seconds. When the user clicks a macrobutton on the client-user interface, the motion is initiated within 1 second or less. The delay most apparent to the user is the 1 to 30 seconds needed for completion of the Cu-SeeMe video update at full resolution. The update is faster for a viewer at the remote site. Little or no degradation in response time was when we switched between a private T1 link and the open internet.

The minirobot is theoretically capable of moving from one joint extreme to the other in under 200 ms [4]. Movements this fast were not demanded, but robot motion appeared smooth, stable, and coordinated at all times. This was also true for small and slow movements observed under the microscope. For example, a 1.0 mm movement of the pipette tip planned for 10 seconds duration was observed under high magnification of the pipette tip. This movement appeared smooth, stable, and of the correct amplitude without overshoot. Periods of acceleration and deceleration consistent with the third order polynomial trajectory generation algorithm were clearly observed.

In these demonstrations, a 0.5 mm plastic cube was substituted for the protein crystal. In one set of experimental operations in which the robot in Seattle was controlled from Huntsville, Alabama, the crystal was successfully captured in three out of four attempts. Each protein capture took about 3 minutes.

18.6   Conclusions

This demonstration system has produced encouraging results which suggest that handling of protein crystals by telerobotics is a viable alternative to the use of scarce astronaut crew time. In addition to the space station application, a remotely operated ground system could allow technicians to avoid difficult micromanipulation tasks performed in cold environments.

There are many opportunities to improve the system. A web interface as pioneered by Goldberg [3] would allow higher resolution video images. A useful feature would be a video streaming protocol for the Internet that supported a user controlled trade-off between update rate and resolution. In these evaluations, users found some occasions in which they needed high-speed video updates and others where they were willing to wait for higher resolution.

Macroprogramming and end-user control would be aided by the provision of Cartesian trajectory planning for the mini-DD robot. Although this type of control was used in the UW mini-DD robot [10], it was not integrated into the PCG test-bed system.

Finally, a more rigorous performance evaluation is desirable. This evaluation should include both task specific and generic experiments. The former will keep development of the system focused on the needs of users and applications, and the latter will provide performance measures that can be compared across systems and applications.

Acknowledgments

The authors are pleased to acknowledge technical collaboration and financial support for this project from Boeing Defense and Space, and fruitful discussions with Dr. Antal Bejczy, Dr. Barry Stoddard, and Dr. Larry DeLucas.

References

[1] H. Asada and T. Kanade. "Design of Direct-Drive Mechanical Arms." *ASME J. Vibration, Acoustics, Stress, and Reliability in Design*, vol. 105, no. 3, pp. 312–316, 1983,

[2] D. Fetterman. "Videoconferencing On-Line: Enhancing Communication Over the Internet." *Educational Researcher*, May 1996

[3] K. Goldberg, M. Mascha, S. Gentner, N. Rothenberg, C. Sutter, and J. Wiegley. "Desktop teleoperation via the World Wide Web." *Proceedings IEEE International Conference on Robotics and Automation*, vol. 1, pp. 654–9, Nagoya, Japan, 21–27 May, 1995.

[4] B. Hannaford, P.H. Marbot, M. Moreyra, and S. Venema. "A 5-Axis Mini Direct Drive Robot for Time Delayed Teleoperation." *Proc. Intelligent Robots and Systems (IROS 94)*, vol. 1, pp. 555–562, Munich, September 1994.

[5] B. Hannaford, A. K. Bejczy, P. Buttolo, M. Moreyra, and S. Venema. "Mini-Teleoperation Technology for Space Research." *Proceedings Micro Systems, Intelligent Materials and Robots (MIMR-95)*, pp. 524–527, Sendai, Japan, September, 1995.

[6] B. Hannaford, P. H. Marbot, P. Buttolo, M. Moreyra, and S. Venema. "Scaling Properties of Direct Drive Serial Arms." *International Journal of Robotics Research*, vol. 15, no. 5, 1996.

[7] B. Hannaford, J. Hewitt, T. Maneewarn, S. Venema, M. Appleby, and R. Ehresman. "Telerobotic Remote Handling of Protein Crystals." *IEEE International Conference on Robotics and Automation*, Albuquerque, NM, April 1997.

[8] B. Hannaford, J. Hewitt, T. Maneewarn, S. Venema, M. Appleby, and R. Ehresman. "Telerobotic Macros for Remote Handling of Protein Crystals ." *Proceedings Intl. Conf. on Advanced Robotics, (ICAR97)*, Monterrey, CA, July 1997.

[9] G. Hirzinger, B. Brunner, J. Dietrich, and J. Heindl. "Sensor-Based Robotics–ROTEX and its Telerobotic Features." *IEEE Transactions on Robotics and Automation*, vol. 9, no. 5, pp. 649–663, October 1993.

[10] D. Y. Hwang. "Teleoperation Performance with a Kinematically Redundant Slave Robot." *Ph.D. dissertation*, University of Washington, Department of Electrical Engineering, December 1995.

[11] H. Kazerooni. "Design and Analysis of the Statically Balanced Direct-Drive Manipulator." *IEEE Control Systems Magazine*, vol. 9, no. 2, pp. 30–34, Feb. 1989.

[12] W. S. Kim and A. K. Bejczy. "Demonstration of a high-fidelity predictive/preview display technique for telerobotic servicing in space." *IEEE Trans. Robotics and Automation*, vol. 9, pp. 698–702, Oct. 1993.

[13] D. Normile, "Search for Better Crystals Explores Inner, Outer Space."

Science, vol. 22, pp. 1921–1922, 22 December 1995.

[14] J. K. Ousterhout. *TCL and the TK Toolkit*. Addison-Wesley, 1993.

[15] B. L. Stoddard, R. K. Strong, A. Arrott, and G. K. Farber. "Mir for the Crystallographers' Money." *Nature*, vol. 360, pp. 293–4, 26 November 1992.

[16] B. L. Stoddard, G. K. Farber, and R. K. Strong. "The Facts and Fancy of Microgravity Protein Crystallization." *Biotechnology and Genetic Engineering Reviews*, vol. 11, pp. 57–77, Dec 1993.

[17] J. Sztipanovits, G. Karsai, and H. Franke. "Model-Integrated Program Synthesis Environment." *Proc. of the IEEE Symposium on Engineering of Computer Based Systems,* Friedrichshafen, Germany, March 11–15, 1996.

Index